LUCY AND DAVID

AND THE

GOD EQUATION

Alan McKenzie

Published in Great Britain in 2011 by
Alan McKenzie
PO Box 1617 Bristol BS40 5WH
www.godel-universe.com

British Library Cataloguing in Publication Data
A catalogue record for this book is available
from the British Library

ISBN 978-0-9567649-0-4

Printed by Lightning Source UK Ltd, Chapter House,
Pitfield, Kiln Farm, Milton Keynes, MK11 3LW

For Ros

CHAPTER 1

Friday 1 October

Lucy Darling was late when she walked into the small lecture room clutching her writing pad, pens and a bag of her favourite jumbo-sized peppermint balls, one of which would choke her before the end of the tutorial. The entrance was at the front of the room, already three-quarters full, so that she had to pass in front of the lecturer's desk, mercifully still unattended, although it was five past two. The floor of the room was level, not tiered like the Department's grand, intimidating amphitheatre, an academic arena, its parabolic dish focusing the gaze of every student down onto the lectern, but, even in this more modest setting, she still felt slightly self-conscious as she made her way along the nearest aisle to an empty row of seats at the back.

She wore a black jersey and her tan full skirt slightly above knee-length – she favoured skirts rather than student-uniform jeans – but, as the eyes of some of the male undergraduates followed her to her seat, she wondered, not for the first time, whether she would feel more comfortable conforming with the jeans and T-shirts worn by the few girls in her first-year physics class.

The seats were arranged in six rows with about eight seats in each row, the kind with an integral tablet that swivelled on an arm once you were seated, so that you had something to lean and write on during the lecture. She made her way towards the middle of the back row, sat down, swung the writing tablet into place and arranged her pad, pens and sweets on it.

She hadn't expected to be this late – it seemed that everyone else in the class was already seated – but she had spent too much time just before the tutorial, trying to find out about an inaugural lecture in ten days' time, which she really wanted to attend, unlike the majority of her first-year peers, it seemed.

Surveying the sea of chattering students from her vantage point at the back of the room, she thought she could recognise maybe half of them, and could remember the names of those she had shared coffee with during this first week of lectures at her new university. As she opened her pad, a man who was clearly the lecturer entered and approached the desk in front of the whiteboard, a collective hush spreading in his wake like a Mexican wave across the room.

Lucy was struck by how young he looked. All the lecturers so far that week had appeared at least middle-aged, but he couldn't be more than 25. He placed a single sheet of paper on the desk, scanned the class for a moment and introduced himself.

'Hi, my name's David Lane.' He paused, looked down at the desk and then up again. 'I'm not a lecturer: I'm a third-year postgrad doing a PhD in theoretical physics.' *Right, that would make him about 24, then.* 'Oh, and they pay me a pittance to give you lot a tutorial every Friday.'

He smiled to acknowledge a couple of ironic 'Ahh's' and then asked: 'So, at the end of your first week with us, would you say it's turned out the way you hoped – or maybe feared?'

'Yeah, it's been OK.' This was from a student in the front row whom Lucy had already identified as the budding class spokesman. She recalled that his name was George. George added: 'In fact, it's been so good, I'll probably come back next week,' duly earning the invited laughter as David lifted the sheet of paper from the desk.

'I gather this is a list of problems you've been given to solve, based on the lectures so far,' he said. He raised his eyebrows to make this a question and received some confirmatory nods. He went on: 'Now, we can do this in

2

two ways. We can tackle each of the problems in turn, but we won't get through them all. Or you can tell me which problems are bugging you most and we'll just work on those.'

'Let's just do the difficult problems,' said somebody, and there were murmurs of support from several parts of the room.

This was fine with Lucy, too. She had been looking forward to this tutorial because it would be her first indication of how she was measuring up to the rest of the class. She hadn't found the list of problems difficult – on the contrary, she had bulldozed her way through them – but she had no way of knowing how easy the others had found them.

It wasn't that she lacked belief in herself, exactly: indeed, she had consistently been top in physics and maths at school. The reality was, though, that any number of these new undergraduates could have been top in *their* school, too. So, for the time being at any rate, she had decided to keep a low profile until she could feel more confident about joining in.

David turned to the whiteboard behind him. It bore the battle scars of an earlier tutorial, the blue equations scrawled across it partially wiped out, no longer universal truths. He began to wipe the board clean and called over his shoulder, 'Somebody give me a problem, then.'

There was a cry for 'number four', followed by more for 'number seven'.

'OK, we'll start with number seven,' said David. The problem was to find the force between a small electrically charged object and an adjacent wire that was also electrically charged. When he had a clear working space, David turned back to the class and asked for proposals on how to proceed. A few hands went up, David chose one, and started to work out the problem on the board using the suggested line of attack. This process carried on for ten minutes, David turning to the students for help at each step of the way.

After several iterations of question and answer, the whiteboard was covered in diagrams, with integral signs and trigonometric symbols nestling between the drawings like mortar binding bricks. By now, the students were fully engaged, and there was no shortage of proposals for the subsequent problems.

Although Lucy had not contributed to the discussions, she had listened attentively to the exchanges. The responses from the undergraduates who had participated were appropriate, accurate and delivered without prevarication, doing nothing to bolster her own self-assurance.

Four problems on, and with the wall clock above the whiteboard showing a quarter to three, she was feeling vaguely sorry for herself. She knew this was an over-reaction – her Dad would have called her a drama-queen – but this wasn't how she'd hoped the tutorial would turn out. True, she had expected her fellow students to be bright, but in one scenario that she had secretly fantasized, hardly daring to play it over even to herself, she had been the only one to come up with the answers to the problem sheet, attracting in this scenario admiring and envious glances from the rest of the class.

That was clearly not going to happen now, and her attention began to stray from the current thread of discussion, focusing instead onto something that had occurred to her during the previous question. The problem had been to calculate the attractive force between two parallel wires each carrying a current in the same direction. The force, of course, was due to the mutual magnetic field between the two wires, but it now struck Lucy that, if the current had been just two streams of electrons in the vacuum of space, without the wires to keep the electron streams on track, then the two streams would repel each other violently.

4

As she played with this thought-experiment, it dawned on her that it could be boiled down to just two electrons travelling side by side. Presumably, if they travelled faster and faster, the magnetic field that each generated would eventually become so strong that it would just balance the force pushing the two electrons apart. So what speed would that happen at, she wondered? Did she have enough physics to work it out? Preoccupied, she rummaged in the paper bag and quietly undid the wrapper of one of her sweets, covering it with her hand to muffle the crinkly cellophane, her movements fluid with long practice, and popped it into her mouth.

She was still toying with her problem when David's voice brought her back to the tutorial. 'OK, I've stuck to the question sheet up to now,' he said. 'In the last ten minutes, is there anything in the lectures that's been bothering you or you'd like to ask about – maybe something that's not in the question sheet?'

Lucy was sorely tempted to blurt out her question about the two electrons. Would that look like attention-seeking, just asking the question to show how smart she was? Or worse, would another student come up with the answer and make her look foolish? On the other hand, she felt instinctively that hers was a deep question. She knew that, if she didn't ask it now, then she would later regret having kept quiet – been faint-hearted, even. So, before she could reconsider, she raised her hand.

David saw her: 'Yes, good, at the back there – what's your question?'

She thrust the sweet into the pouch of her cheek as best she could and raised her voice: 'Well, I was thinking about that problem, the one before last, and I wondered about just two electrons travelling side by side.' The sweet made her feel like a hamster, and she was sure her voice sounded strangulated, but she had started now, and so she had to go

5

on. 'Of course, I know they would repel each other, but there would be a magnetic attraction between them as well.'

'Yes, that's right,' said David.

'Well, I was just wondering – how fast would they have to go before the magnetic attraction just balanced the electrostatic repulsion? I mean, so they didn't fly apart – do you see what I mean?'

'Yes, I see exactly what you mean,' David replied slowly, as though playing for time while searching for the answer. 'Well, what a fascinating question!'

To Lucy's relief, he didn't immediately offer it as a problem for the rest of the class but, instead, turned to the board and began to write an equation on it. He looked as though he was explaining it to himself rather than to the audience. As he wrote, he spoke the symbols softly, distractedly, clearly working this out for the very first time.

Suddenly he hit the board with the blue felt-tip pen, creating a violent full-stop. 'Of course!' he exclaimed. He turned round and faced the class, faced Lucy: 'What a lovely problem! The answer is that they have to go at the speed of light! When these two electrons reach the speed of light, then the electric field and the magnetic field will exactly cancel.'

Lucy was thrilled. She couldn't prevent a smile from bursting out, and she fought to control it. But the muscles of her smile squeezed on the huge peppermint ball where it was being held captive between her upper and lower molars. With no warning, the sweet catapulted straight down her throat where it lodged – horrifyingly – in her larynx.

Her immediate reflex was to cough, but the involuntary inspiration preceding the cough only sucked the sweet further down into the unyielding trachea, plugging it completely.

David continued, looking round the class, his excitement palpable: 'This result shows the deep connection between the electrostatic field and the magnetic field. In fact – you won't learn this until next year, maybe the year after – but you can explain the magnetic field completely in terms of the electric field if you use special relativity. That's a beautiful result!'

Lucy's whole world now was focused on her throat. She tried to breathe in but her airway was solidly blocked. She could feel panic rising as she thumped herself, first high on the breast-bone and then higher, on the front of her throat where she could feel the obstruction.

This did nothing to remove the blockage. She could feel the sweet move up and down maybe a centimetre or so as her lungs worked futilely to exhale and inspire, but no air was getting past the plug.

There was a roaring in her ears and she sensed people turning round in their seats to look at her. Not fully aware of what she was doing, she shot up out of her seat, upending the writing tablet so that all her things fell to the floor. Half blindly, she crashed along to the end of the row and ran down the aisle clutching at her throat with both hands.

She could hear David call to her, muffled and distant: 'Are you choking?' but she no longer had the presence of mind to respond. She flayed out her arms as he caught her and spun her round to face the class. She could feel him thumping her back as her field of vision began to darken and contract. She sagged forward, sucked slowly and inexorably down into the throat of a black whirlpool.

The roar was becoming distant now, and she was reminded in a detached way of languishing gracefully under water, far from the cares of the noisy, busy world above. She felt as though she were an observer, watching this happen to somebody else, a still-functioning corner of her mind uttering the phrase *out-of-body experience*, the very utterance as random, unrelated, disconnected from her as her body itself was now. She noticed unconcernedly the arms that were clamping her, compressing her, the fists that were thrusting up into her diaphragm. Once. Twice. Three times.

Time is slowing down.
That's because I'm dying.
Lucy, I've been waiting for you!
Dad! I've missed you so much!
Blessed darkness...

With an explosive cough the sweet shot out of her mouth and she burst up through the surface of the whirlpool. God's sweet air rushed into her lungs as she took a loud, ragged breath, then another. The roar in her ears came rushing back – the class were cheering, clapping, whooping!

She was still in David's arms, and he gently guided her to a seat at the front – the class had largely vacated their seats and were standing around her and David like spectators at an impromptu boxing match, shouting and laughing now, not at her, but with a hysterical edge, their adrenalin seeking an outlet, George, the class spokesman, holding out a red-and-white striped sphere, dust and hair sticking to it like grey dandruff: 'You dropped this...' followed by renewed shrieks of laughter.

Lucy could feel her eyes brimming with relief. David seemed to be the only one not laughing – he hunkered down beside her so that his face was level with hers, concern furrowing his brow. 'Are you OK?' he asked her.

She nodded. 'Need to catch my breath—'

David looked up at the crowd around him, and raised his voice. 'All right,' he said, 'tutorial's over for today. Let's give her some space.' He glanced back at Lucy. 'I'm going to take her along to the Community Hospital when she's recovered enough. I'll see you all here again, same time, next week.'

The class gradually filed out, many of them stopping to wish Lucy 'good luck' and 'all the best', with a few cries of 'well done' directed towards David.

When the last student had left the room, Lucy had improved enough to speak normally: 'Look, what you said about going to the doctor's – I don't need to go; I'm absolutely fine now.'

'How about your throat? Does it hurt?' David had shifted into the seat beside her.

'Only just a little. I'm a bit tender here' – she put the flat of her hand on her front at the bottom of her rib cage – 'but nothing alarming. Was it you who was squeezing me?'

'Yes, I'm sorry if it was too hard. I tried a few times and nothing happened, so I kept increasing the pressure each time. I guess I was getting a bit desperate.'

'Not half as much as I was! Thank goodness you knew what to do!'

'Well, I didn't really. I mean, I've never been taught it – I've only seen it done in films. I'm just so glad it worked.'

'Me too!' She smiled at David. 'You know, I never thought I would ever be in a position to say this to anybody, but you saved my life! Thank you!'

David looked down at his shoes, clearly embarrassed. 'Oh, no, if it hadn't been me, remember there was a whole classroom full of students here.' He rose out of the seat and turned towards her. 'Look, if you really don't want to go to the doctor, what are you doing next? Do you have a lecture?'

'No, that's it for the week. Nothing till Monday.'

'Then why don't we go for a coffee or something? I'd be happier if I could just see that you're all right.'

'What about you? Don't you have any more lectures?'

'No, and I'm not a lecturer, remember, just a postgraduate student. I've nothing till a meeting at four o'clock. Will you come?'

Lucy smiled and stood up. 'OK, that would be good. And maybe I can ask you a bit more about that problem? I think I missed the last bit of what you were saying!'

'Just a minute!' David strode back up the aisle, worked his way along to where Lucy had been sitting, and retrieved her notepad, pens and bag of sweets from the floor. He handed her things to her as they went out the door.

'Thank you, Mr Lane.'

'Oh, goodness, no, it's David. Please call me David. Everybody's on first-name terms at university. But I don't know *your* name,' he added.

'It's Lucy. Lucy Darling.'

'Lucy – *light* – that's appropriate, somehow, under the circumstances.' They were walking along the corridor, now, side by side. 'It comes from the Latin, but I expect you know that,' he added.

Lucy appraised him from out of the corner of her eye as they arrived at the entrance hall of the Physics Building. He was as much as six inches taller than she was – six-two, maybe even six-three. He had straight, dark-brown hair which had an engaging tendency to flop over his forehead so that he turned his head slightly to see past it when he looked at you. Somehow, it added an attractive note of vulnerability, for all his height.

They emerged into an unseasonably warm and bright October sun. Lucy turned to David and said, 'Shall we go to the New Hall coffee bar – it's probably the nearest one?'

'Good idea. I'd heard they had a café – is that your hall of residence, then?'

'Yes, it's almost like a hotel! We all have en-suites, and the hall can provide all your meals, so you don't waste time cooking if you don't want to. There's a flat-screen TV in every room – of course, they're really catering for the conference and holiday season out of term-time.' She wondered whether to ask David where he lived, but that didn't feel right; it would be too familiar. Despite his protestations about being a student, he was still six years older than she, and a tutor, to boot.

'I lived in halls when I was an undergraduate,' said David, 'but now I have a place in Spey Street.' He turned to look at her. 'How long have you been interested in physics, Lucy?'

'When I was quite young, when I was in primary school, I was fascinated by space research. I was intrigued by how the space station stayed up in orbit, and how you could sometimes see it speeding its way eastwards across the starry sky and how spherical globules of water could float around in a space ship, and I loved the historical footage of men on the moon – that one where the astronaut drops a feather and a hammer and they hit the ground at the same time. Then I found out that the things about space research that really interested me were all part of a science that went by the strange-sounding name of *physics*! When I got to secondary school, I was encouraged by my physics teacher, who could see how enthusiastic I was for his subject.'

She had had a crush on Mr Stevens since her first science lesson in the school. He had continued to take the top physics classes as she progressed through the years, her intrinsic interest in physics, her aptitude and his teaching style all combining to fire her motivation and determination to shine in the subject.

She recalled to herself how she would stay late after his classes to clean the blackboard, always putting up her hand eagerly to answer his questions, borrowing a friend's pencil case and then knocking on his classroom door to return it, just to see him. Looking back, she realised that he must have been well aware of the crush, but he had handled it kindly and sensitively.

A high point in her campaign had been the experiment with the Van de Graff generator. Mr Stevens had arranged the kids into a daisy chain holding hands, with the student at one end of the chain keeping his hand flat on the dome while the machine was cranked up. Mr Stevens positioned himself at the other end of the chain and then

11

grounded himself and the whole chain by reaching out to touch a water tap. The jolt that had gone through all of the kids was literally shocking, and left them with a lasting impression. Lucy's strongest memory of the experiment, though, had been the fact that, somehow, she had managed to arrange it so that it was her hand that Mr Stevens had held during the experiment...

None of this, of course, she told to David. Instead, she said, 'I can't imagine myself doing anything other than physics. Two of the students in my last year went to medical school, and one got into vet school. Others went into music and languages and that sort of thing.'

'Why do you think it is that relatively few women choose to do physics?' he asked (she liked that he had said *women*). 'There are only three – four, counting yourself – in your class.'

'Maybe it's the boys who aren't rounded enough! If the boys in the physics class were equally interested in the arts, then maybe some of them would have chosen to do languages or English Lit instead.'

This started a debate that took them right up to New Hall, where Lucy led the way to the coffee bar with a new resident's proprietorial pride.

'Would you like your caffeine ice-cold?' David asked her at the bar. 'I mean, would you like Coke – maybe it would be better on your throat than hot coffee?'

'Yes, actually that sounds like a good idea. Could I have regular, please?'

'Great, grab a seat and I'll bring them over.'

Lucy found a table with two seats opposite each other and sat down on one, watching him making his purchases and carrying them over to her table.

'I thought we'd have a Coca-Cola float,' he said, setting down a tray with bottles, glasses and two bowls of ice cream. 'Not had one before? Watch this!' and he scooped

12

the ice cream into each of the glasses, poured the Coke over the ice cream and stirred.

Lucy took hers and spooned some of the frothy ice cream into her mouth. She couldn't resist smiling as it slid down her throat, soothing it and quenching the embers of her recent trauma.

'It's surprising how good that feels, considering I'm not cooling my airway directly.'

David looked pleased. 'You know, that was a remarkable question of yours. That was the kind of question that Einstein asked himself and led him onto special relativity.' He took a spoonful of float and said: 'You waited until the end of the tutorial – why didn't you join in the discussions earlier? You must have known the answers!'

'I was trying to keep a low profile.'

For a heartbeat, David just looked blankly at Lucy while the words sank in and then the corners of his mouth twitched. Lucy could feel the giggles burbling up inside her and she let it come, joined by David, and together they allowed the floods of healing laughter to drain away at long last the pent-up tension of their shared ordeal.

'Oh my, I'm sorry,' David was finally able to say, as he dabbed at his eyes with his paper napkin. 'Oh, but I so needed that!'

Lucy patted her own eyes with her napkin – *I'm lucky to be alive at this moment!* 'At the risk of starting us up again,' she said, 'I probably should explain what I meant by wanting to keep a low profile.'

David grinned. 'Yes, I'm very curious.'

So Lucy told him about her anxiety that being top at school could nevertheless mean being only average at university. It was this uncertainty that had held her back from taking part in the tutorial in case she made a fool of herself.

'Funny enough,' said David, 'I remember wondering the same thing about myself. It wasn't until the first-semester exams that I knew where I ranked in the class.' He lifted the glass to his lips, leaving a white moustache. 'But that question you asked shows you have an intuitive feeling for physics – my guess is you'll find yourself near the top, if not *at* the top.'

Lucy could feel herself blushing at his praise, but she spoke as nonchalantly as she could manage: 'OK, maybe I asked the right question, but, to be honest, I still don't understand why the magnetic force becomes as strong as the electric force at the velocity of light. I missed what you said after that.'

'No, I didn't explain it. One way to see it is to work out how forces change at speeds close to light. It's part of special relativity.' He wiped his lip with his napkin. 'Look, can I borrow your pad?'

He moved his seat round beside Lucy and for the next twenty minutes he showed her, step by step, how the magnetic field was, in fact, just the electric field transformed by relativistic effects. Lucy was thrilled.

David picked up on her excitement. 'You know, you're very good for my ego,' he said. 'I'd like to flatter myself that my teaching style appeals to you, but that's not it, is it? You're really into this stuff for its own sake, aren't you?'

'To be absolutely honest, I'm enthralled by the ideas in relativity. I mean, yes, I love the way you explain this, but you're right, there's more to it.' Recollecting, she peered into her empty ice-cream bowl, rotating it absently by the stem. 'It was when I was in my early teens that I first found out about relativity. I was hooked! Of course, it was the weird stuff with time that really gripped me.'

David nodded, not interrupting.

'Well, I managed to teach myself the basics of special relativity from the internet.'

'What – without anyone helping you? Or guiding you?'

'Lord, no! I did all that in secret! I was already regarded as a freak – my friends all thought I was a swot, which

14

wasn't entirely fair, because the only private study I was doing was stuff that wasn't in the school syllabus.' It had been a fine line to tread, balancing her enthusiasm for physics with the risk of being ostracised by her friends.

David smiled. 'So now it's a relief – liberating – to talk about relativity with another person without feeling guilty or worrying about being called a swot?'

He understood! 'Yes, you've got it exactly!' she said, returning his smile. 'Is that what happened to you, too? Did you teach yourself physics?'

It was David's turn to reflect. 'I suppose,' he said, after a pause, 'that you could say my interest in physics began with science fiction. My father had a large collection of science fiction and I started to bookworm my way through it in my early teens. Later, I moved onto physics books.'

'Science fiction! That was my passion too, even before relativity! You can't beat the greats – Isaac Asimov, Arthur C Clarke, Robert Heinlein—'

David broke in, 'Yes, did you ever read Heinlein's *Starman Jones*?' Lucy nodded and started to reply but he was unstoppable, eyes shiny with enthusiastic nostalgia.

'It was written for kids and it was the very first science fiction book I ever read. You know, he talked about crossing the universe using warped space and mentioned the possibility of parallel universes in that story long before these ideas became fashionable. And there's a paragraph where he quotes *pi* to 29 decimal places and I learned them all and I even used to recite them to myself when I was brushing my teeth!'

Lucy laughed. 'Do you still remember them?'

'Oh, sure, in fact, I can do *pi* to 41 places now, but I'm not going to or you'd have your proof that I'm an obsessive neurotic!'

'What about older science fiction? Did you read H G Wells' *The Time Machine*?'

'I read it after I saw the film – the original film, I mean. The second film wasn't nearly so true to the book.'

'Yes, you're right,' Lucy agreed. 'But there's a very poignant passage in the book that isn't in either film, when the Time Traveller accidentally pushes the lever forward in time, and he travels far, far into the distant future. So far, in fact, that the Earth has synchronised its daily spin with its solar orbit so that it always points one side towards the sun, just as the tidal forces raised on the moon by the Earth eventually forced the moon always to point one side towards the Earth.'

She paused, but David was just gazing at her, not about to say anything.

'When I first read that passage, I cried. I thought how far away in distant time all of his family and friends must be, so long gone, so long dead. And the Earth was dying too. It made you realise that we're still in the rosy summer of the Earth's time, but that there will not always be green, grassy hill tops, blue skies and wispy clouds floating in a gentle breeze. No matter how many cycles of climate change we go through, there will be a *last perfect day* on Earth, as Carl Sagan put it, and, after that, life will gradually become intolerable as the sun moves into its red-giant death throes'

Suddenly she was self-conscious. *What must he think of her, opening herself up like that?* She was about to lighten the mood when David took up her train of thought.

'You know, it's strange,' he said carefully, 'to see you sad about things that won't happen for billions of years. Most people would say they wouldn't even be concerned about events a mere thousand years from now. But actually, I share your sadness. It makes me quite disheartened to realise that, no matter what we do, life on Earth will not go on forever.'

'Yes, I don't suppose it's rational, is it? I've tried to understand why I feel this way. Basically, I suppose, it's not so much sadness for loss of the Earth; it's more sadness for loss of the human race.'

'Maybe most people don't really think that Earth or the human race will ever end.'

Lucy nodded. 'I think that's the point. I mean, they believe it intellectually, of course, but not with the heart. It's like looking at the Swiss Alps and being told they're continually being formed by continental plates colliding and pushing rocks upwards like folds in a tablecloth. But you never see this happening. It's far too slow. People build chalets and live for generations in the Alps. And it's the same with the whole Earth, on a grander scale.'

David looked down at the table, deep in thought. 'Somehow,' he said, 'we think it's going to go on forever because, in our short lifetime, we haven't *experienced* changes on a geological timescale. Just as we can't really imagine the world after we're dead, because as soon as we try, we're still there, in our imagination, looking down, ethereally, from the corner of the room or whatever. Whereas, in reality, our sense of existence will stop when we die.'

Lucy wondered whether to put her next question. It was so personal, but she felt that David wouldn't mind being asked, and he had piqued her curiosity. 'Does that mean you don't believe in God?' she asked.

David glanced at Lucy then looked down at the table. 'Ah, I should have seen that coming,' he said. 'I'm always wary of upsetting people in this situation.'

Lucy was puzzled. 'What do you mean, *this situation*?'

David was hesitant. 'Well, from the way you asked your question, I'm guessing that you *do* believe in a god, and, well, the fact of the matter is that I don't, I'm afraid.' He looked up at her again, and Lucy was amused to see that he actually looked guilty.

'Why be *afraid*? Surely you're as entitled to your view as I am to mine?'

17

'Yes, of course,' said David, 'but my beliefs have landed me in trouble before. You see, I'm firmly in the atheist camp, and I used to preach, if that's the right word, without considering whether I was offending anyone.'

'So, what you're really saying is that it was the way you defended your beliefs, rather than the beliefs themselves, that got you into trouble!'

David smiled wryly. '*Touché!* So – are you a Christian, then?'

Lucy considered how she should answer this. She had been brought up quite liberally by parents who attended church fairly regularly without insisting that she follow their example. Gradually, throughout her teens, she had increasingly questioned what she perceived as the dogma upon which Christianity was based and it now seemed to her that the only backing for the Christian story was contained within the Christian story itself. That was a circular argument and she could think of no independent corroborating evidence.

However, she would feel guilty about admitting this out loud, especially to a stranger. She felt that she would in some way be letting her parents down if she did so, particularly because she had never told them of her developing scepticism, although they may have suspected. On the other hand, she was an adult now, she was independent, and she should be prepared to voice her beliefs. David would be a good person to confide in – he was mature, sympathetic, and, she had to admit, quite good looking!

She gave in. 'You could say that I follow the Christian philosophy but not the theology. By that, I mean that I respect others as equals and individuals, and all that follows from that.'

'So you don't believe in a god then?'

'Ah, I was coming to that. I don't believe in a god in which there are three persons, the Father, the Son and the

18

Holy Spirit; in fact, I never understood what that even meant! Of course, I do believe in the historical fact that Jesus Christ existed – there are too many independent historical accounts for that not to be true. But I don't believe that he was a god.'

Two students came in and sat at the next table, and Lucy lowered her voice. 'But I'm not an atheist either – whatever explanation is put forward to explain the existence of the universe, you can always ask – *but what caused that?* The only way you can stop a sequence of such questions becoming infinite is if the final answer is God.'

David looked intently at Lucy and she noticed for the first time how blue his eyes were. 'Lucy,' he said earnestly, 'if you think what I'm about to say sounds condescending, it's absolutely not meant to be. I think that's the most rational and appealingly simple argument for a god that I've ever heard. If I weren't a consummate atheist, you'd be in danger of convincing me!'

'Well, thank you – I think!' To get more comfortable, she half-turned her chair towards him and propped her chin in the palm of her hand, her elbow resting on the table. 'So let me get this right. As an atheist, it's not a question of not believing in a god – you actually *know* there isn't a god?'

'Yes, that's not a bad summary of how I see it. *Believing* is for the other side!'

'OK. Then, as a scientist, a physicist, you must have backed up that knowledge with a pretty convincing proof that there is no god!'

David grinned broadly, crinkling the corners of his eyes. 'Ah! Now we come to it!' He was evidently excited. 'Leaving aside the question of what is meant by a god – and that's a pretty central question – by coincidence, and it really is a coincidence, that turns out to be close to what I've been working on for the past six months!'

Lucy put her hand flat on the table and straightened herself up, surprised. 'What, for your PhD, you mean?'

'No, no, nothing like that. Although, funny enough, my PhD topic does have a distant connection.' Abruptly, he checked his watch. 'Damn, I've got to go,' he said apologetically. 'How are you feeling?'

'I'm fine, now, really. And thank you for looking after me. And for saving my life!'

David stood up. 'I'm just so glad it turned out all right. If you have any problems at all, or you're the least bit worried, will you promise me that you'll call a doctor?'

'Yes, of course. And maybe we could continue our debate sometime? I'm really curious to know what your proof is!'

'Ah, I didn't say I had a proof. In fact, the very question of proof is what I've been working on.' He leant over towards Lucy to keep the conversation private. 'Without meaning this to sound too melodramatic, I think of what I'm doing as trying to solve the riddle of the universe!'

He started to retreat and then stopped. 'See you next Friday?' he asked.

'Yes, I'll look forward to that very much,' said Lucy and waved goodbye. She watched him as he left the bar. Had he meant that he would just see her in the Friday tutorial, or did he mean that he would meet her afterwards, like today?

Her mind was buzzing with the experiences of the last two hours. First, she had nearly died. Then she had been rescued in the most dramatic manner by a tall, good-looking, intelligent, mature and sensitive young man who shared her interest in physics, relativity and science fiction! Was he single? With those credentials? Dream on!

What was she thinking! Ever since she had known she was coming to university, she had been aware that she would meet eligible young men. She had resolved, though, that her guiding principle should be to work to her full potential in her chosen subject, physics, and, among other things, that meant not being distracted by a full-time relationship. Ideally, she would find that she could cope with the work and still have time left over to enjoy herself. However, even in the best-case projection, she wouldn't be sure of how she was doing until the end-of-semester exams, and, more likely, the end-of-year exams. Until then, she must concentrate on doing as well as she could.

In any case, she reasoned, David was just being polite in saying he would see her next Friday. He had asked her to the coffee bar not because he was attracted to her but to be sure that she wasn't any the worse for her choking spasm or even injured by the abdominal thrusts that he had performed on her. He was six years her senior – he must regard her as virtually a school girl, although, to be fair, everybody had always said she was mature for her years, which she tended to agree with.

She had never had any boyfriends from within her own class, or even her own year, for that matter. Her girl friends had told her that she intimidated the boys – probably a mixture of her academic prowess (in physics and maths, at least) and the fact that (yes, it was OK to admit it to herself, in private) she was actually quite good-looking. She had long, thick, straight chestnut-brown hair, a nose that you might describe as pert and, mercifully, skin that had not been visited by the ravages of acne, from which she was one of a very small group in her class that seemed to have escaped entirely.

When she was 14, she had gone out with a boy two years older, but the relationship had been essentially platonic. They had gone to dances, movies and parties, but nothing had *happened*, for which she was grateful. The romance had lasted, amazingly, for two years, until the boy left school for university. Being paired with such a senior school student had probably done nothing to allay the intimidation felt by her peers, she reflected wryly.

So, she should put all thoughts of David Lane to the back of her mind and spend the weekend going through her lecture notes so that she would be ready for the following week. The thought flashed across her mind that she was glad, after all, that she hadn't worn the obligatory jeans and top – she had been dressed her casual best for David. And then she thought what a spectacle she must have made in front of the class, being man-handled like that. But to think that she had been encircled within his arms...

At that moment, her hall friends, Gillian and Sarah, came into the bar. She was so glad to see them: she badly needed a long chat...

CHAPTER 2

Six months earlier

Six months earlier, there had been no thought in David's mind about solving any riddles of the universe as he set off late one evening in April for Mike's party. Like many postgraduate students, David lived slightly out of phase with most of society, rising not much before ten in the morning, going to lunch around two in the afternoon, not eating his evening meal until late and going to bed late, usually after midnight. This wasn't a deliberately planned cycle: it arose essentially because his most productive, creative hours were after dark, so that he tended to work well into the small hours. Consequently, he slept until later, and this re-set his biological clock to mid-Atlantic time.

It was therefore nearly 11 pm when David arrived at the party. He had known Mike since undergraduate days at the same university, and he now shared a postgrad office with him in the Physics Building, although Mike was an experimentalist, not a theoretician like himself. Mike had recently moved into a rented flat with his girlfriend and they had decided to throw a flat-warming party. Judging by the level of noise when he entered the flat, he was already several drinks behind. He edged his way after Mike into the small kitchen, dumped a six-pack into the last space left on the drainer, helped himself to a clear plastic cup and filled it from a beer barrel on the fridge.

'I think you know everyone here,' said Mike, looking back as he left the kitchenette and started to edge through the crowd towards the door once more. With five people including David, the kitchenette was full. Two were the girlfriends of two physics postgraduates currently in the living room, there was Bill, another theoretical physics postgrad student, and the fourth person, whom David had not, in fact, met before, was introduced as a *post-doc*, Peter Brown, who was in the middle of a two-year post-doctoral research fellowship, having completed his PhD in pure mathematics.

They had evidently been discussing the perennial two-cultures question. 'What makes us different from the humanities people,' Bill was saying, glancing at the two women, 'is the scientific method. I mean, we scientists try to make sense of things by thinking of the simplest explanations for things – events – that we see around us. Well, the scientific method would say: think of an explanation for what you see, and, if it's a good explanation, you will be able to extend it until it predicts something you haven't seen yet, and, if you then look for that thing and you find it, well, that's evidence that your explanation fits reality at least to that extent. If it's not there, though, then you have to modify your explanation or ditch it altogether.'

Helen, one of the two women who had been waiting impatiently to make a point, broke in. 'Sorry, but what's so exclusive about the scientific method? Are you saying that only scientists use it? In that case, a whole lot of us who thought we were in the humanities are clearly scientists! And anyway, surely even scientists don't use the scientific method every minute of the day. I mean, don't they sometimes just *observe* without theorising?'

'Yes, but Bill also said that we look for the simplest explanation,' said David, joining in the conversation, 'and that's characteristic of the scientific method. That's even been elevated to a principle—'

'Occam's razor, yes.' Peter the post-doc spoke for the first time. His voice had a deep, quiet assurance that befitted his post-doctoral status. Of all of the people in the room, he had actually gained his PhD – he was entitled to be called *doctor*. His large physical stature added to the gravitas lent by his seniority amongst those present, and his movements were unhurried as he paused to tamp down the tobacco in his pipe. Whether this was for effect or whether he was genuinely addicted to the weed was unclear, but, either way, they kept respectfully quiet as they waited for him to elaborate. 'Actually,' he continued, 'I think it would be a mistake to think of Occam's razor purely as a principle of favouring the simplest explanation.'

'Sorry, Occam's razor?' queried Helen.

'I was coming to that,' he said affably. 'You can think of Occam's razor as a razor paring away all unnecessary additions to a hypothesis until you're left with the bare bones that still work. Oh, Occam, by the way, was an English monk or a philosopher or something like that, probably back in the Middle Ages. Let me see if I can think of an example.' He sucked tentatively at his pipe. 'Yes, suppose I have three different measurements of something and I plot them on graph paper and I find I can draw a straight line through them. Thing is, I could also draw any old curve as complicated as you like on the graph paper and I could make that curve also go through all three points. Occam's razor would eliminate all these additional curves and swirls until we were left with the straight line. Unless I have any other information to the contrary, I then assume that my measurements are determined by a law that produces a straight line.'

'Although you could argue,' said Bill, 'that the straight line is the simplest solution, so Occam's razor really does favour the simplest explanation.'

'Well, it depends on how you define *simple*, I guess,' replied Peter, 'but you can see how Occam's razor works within the scientific method, because, if it turned out that the next measurement was no longer on the straight line, we'd be forced to look at the next most economical line, which might be a regular wave, say.'

Helen came in mischievously: 'So you could use Occam's razor to say that the simplest explanation for how the universe was created is God,' she smiled. 'Far simpler than all the stuff about the Big Bang and spacetime foam and all that!'

'For an English teacher claiming not to be a scientist, you certainly know your science jargon!' David observed.

'Comes of having a theoretical physicist for a boyfriend!'

'Ah, now we come back to the scientific method,' said Peter. 'You see, for many scientists, a hypothesis isn't respectable unless it contains tests that would allow you to decide whether it is a good representation of reality. If your definition of God is simply the thing that created the universe, then nobody can disprove your hypothesis that God created the universe, because it's a circular argument. But if your god is also a thinking being that has concerns for humanity, you could equally well use Occam's razor to show that it is unlikely that your god created the universe.'

'How come?' asked Liz, the other woman in the kitchenette.

'Well, the origin of the characteristics of the universe that we observe can be explained fairly simply in quantum mechanical terms,' said Peter, 'and so to introduce a thinking being into the explanation is unnecessary. Like adding curves where the straight line would do.'

Bill was evidently still bothered about the differences between the sciences and the humanities. 'Talking of creating,' he said, 'one of the misunderstandings between scientists and non-scientists is that the non-scientists say that they are the only ones who create things, like Shakespeare did, whereas scientists only discover things that were there already.'

26

'I don't say that,' objected Helen, mildly, sipping her red wine.

'Maybe not, but it *is* a fairly common misconception,' said Peter. 'There's no doubt that scientists and mathematicians *do* create, and they create according to sets of rules just as Shakespeare did when he wrote his sonnets according to a set of rules. And the mathematics that they create may well have nothing to do with any known physical reality, so they can't be said to be discovering things that were there already. Not uncommonly, mathematicians will invent a framework according to a set of rules they invent just to see where the rules lead them, and then, much later on, a physicist will come along with some observations and will spot one of these mathematical frameworks that fits the observations and so he adopts it. Non-commutative algebra is an example.'

'—or she!' said Helen.

Peter continued: 'Bertrand Russell was eminently a creator. Have you heard of Russell's paradox?'

None of the company had. 'OK. In a Spanish village – I don't know why it's in Spain, and it really doesn't matter where it is, so bear with me – in a Spanish village, there is a barber. He shaves all of the men who don't shave themselves, and he doesn't shave any of the men who shave themselves. So – does the barber shave himself?'

They considered this. 'How could we possibly know?' asked Liz.

'OK, well suppose he does shave himself,' suggested Peter. 'Would that be an acceptable answer?'

'Ah, I get it,' said David. 'If he shaves himself, fine, but you said that the barber doesn't shave any of the men who shave themselves.'

Peter smiled. 'Go on,' he encouraged.

'Well, since the barber shaves himself, he must be one of the men that the barber doesn't shave. So the barber doesn't shave himself, but we started by saying that the barber does shave himself!'

'Ah, yes,' exclaimed Helen, 'and if he doesn't shave himself, then, because Peter said the barber shaves all the men who don't shave themselves, then the barber must do it. The barber shaves himself after all!'

'You've got the paradox,' said Peter with pleasure. 'But actually, the fact that this paradox could be stated in the mathematical system that Russell was developing, was eventually to bring his life's work virtually crumbling around his ears!'

They waited expectantly while Peter paused again to re-ignite his pipe. He wore an old-fashioned tweed sports jacket, with leather pads at the elbows. The pockets bulged, presumably with the paraphernalia that he conjured up from them to maintain his pipe in good working order. His tortoise-shell glasses, the concavity of the lenses indicating moderate myopia, completed the image of a stereotypical academic, which, of course, he was.

'Actually, I suppose that may be somewhat melodramatic,' he admitted, 'but grant a mathematician some artistic licence! You see, Russell, and some others around the nineteen hundreds, were trying to create a perfect system of mathematics. They were in the middle of doing this when Russell discovered his paradox. This was a blow, because the paradox could be written down using their mathematical system, which meant that their system contained inconsistencies. So the system wasn't perfect after all and so they had to invent patches to hold their system together. The patches were never really satisfactory, although people put up with them until Kurt Gödel upset the whole of mathematics.'

He paused again while he sucked at his pipe. 'You like to tease, don't you!' smiled Liz.

'This time, I'm not exaggerating. Gödel developed a theorem, later called his *Incompleteness Theorem*, which showed that the goal – the dream of all of these mathematicians – the dream would never be realised: there was no pot of gold at the end of the rainbow. You know, I once understood Gödel's Incompleteness Theorem for an hour, and that was a life-changing moment for me. For an hour, I could see the full argument in all its clarity, and then I began to lose small pieces of the jigsaw so that I could still see the basic structure but some of the detail was lost in the holes.'

His audience smiled. Clearly, Peter had uttered these phrases before, practised and polished them in front of other audiences, but they were intrigued none-the-less.

'You're winding us up now, aren't you?' asked Helen. 'I mean, you really understood a theory for an hour and then – what – you *forgot* it?'

'A theorem. Yes, it's difficult to explain, exactly, but it's a bit like reading a good detective novel for the first time. All the clues are there so that when you get to the last page or so, the whole thing suddenly makes sense, if the novel is well constructed. The novel would be incomplete without each of the clues. But if you re-read the book say a year later, you may remember the outline of the plot but you won't recall each of the clues until you come across them again. Well, that's what it's like with Gödel's theorem – you remember the overall structure, but you forget the detail of the proofs that lead up to the punchline. But you know that you can always go back and refresh your memory if necessary, if that makes any sense.'

David was absorbed. 'Yes, I think it does. But you can still see how the theorem works in principle?'

'Oh, yes, of course! In the broad sense, at a high level, it's not difficult to understand at all. Look, I can show you in a sentence – *This statement cannot be proved in mathematical system X.*'

Nobody said anything; they waited attentively.

'Well, what Kurt Gödel did was to show how any sufficiently complex mathematical system – the kind Russell and his friends were trying to create – he showed that any such mathematical system would be able to express that sentence in mathematical language. Now, clearly, the sentence was true, because, if it wasn't, that would mean it *could* be proved in the mathematical system, which would mean the sentence itself is lying. So the mathematical system would be no good – it would contain inconsistencies because you could prove something that lied, that was false, like proving two plus two is five.'

He paused again, but this time, not to attend to his pipe, but to look at each of his audience in turn. He was allowing them to catch up, to get hold of the idea. They returned his gaze, each deep in thought.

'So this mathematical system contains a true statement, in mathematical language, but the system can't prove it. This was the kiss of death to all of Russell's attempts to create a complete mathematical system, because the system would only be complete if you could start from a few universal truths – axioms – and then use these axioms either to prove or disprove any other statement in the system. What Kurt Gödel showed was that all advanced systems contain statements that are true but cannot be proved. That is, assuming they're sophisticated enough to be able to express statements like that one.'

David was the first to break out of their shared reverie. 'OK, but you're assuming that the few axioms you referred to cannot be proved, right? Because they're the starting point. Then why don't you just add one more axiom to the system, which says *this sentence cannot be proved in system X*, and then you wouldn't have an incomplete system, because you've raised the true statement to the status of a universal truth?'

'Good try,' said Peter. 'But then I can always generate a sentence using this, now modified, system *X* to say *this statement cannot be proved in modified system X*, and so we're back where we started. It would go on for ever – you'll never find a system, or a modified system, or a modified, modified system where there are no true statements that cannot be proved in that system.'

Unaccountably, David felt himself kindled by the simplicity and the profundity of what he had just heard, and, sceptic though he was, he felt that he had just been given a glimpse into the future – *his* future – and that he was at one of those pivotal moments when he faced the choice between two life-paths that would continue to diverge into the hazy distance of his time to come.

He made his choice. 'Peter, you make it sound relatively simple, but I'm sure the devil is in the detail. My guess is that it's much harder to translate your simple statement into mathematical language than you make out.'

Peter nodded. 'Oh yes. That was Kurt Gödel's genius. And that's the bit that's so hard to hold in your head for any length of time.'

'In that case, I'd be really interested to see the mathematics of this. How would you recommend I should go about it?'

'No problem,' Peter replied. 'Come and find me in the Mathematics Institute on Monday or any time – they'll tell you where to go at the reception desk. I can lend you a copy of an English translation of Gödel's original paper, and I can also lend you some commentary that you'll need, because Gödel's paper is heavy going until you understand all of the symbols and conventions he uses. It's not for the faint-hearted!'

The rest of the party passed pleasantly enough, but David was preoccupied with what he had just learned and the question of whether he should invest time in understanding the details of Gödel's Incompleteness Theorem. His PhD research was progressing steadily after a slightly shaky start. The topic had been chosen for him, and, while he found it interesting enough, he was not obsessed by it, and so he felt that he could afford to indulge himself a little in the second half of his three years of postgraduate study. If his instincts about Gödel's Incompleteness Theorem turned out to be correct, it could be the most important theorem in the universe – or beyond...

CHAPTER 3

Monday 4 October – Friday 8 October

Lucy looked forward to Monday all weekend. She had done no work for the remainder of Friday afternoon and evening, which had made her feel guilty. She had talked for nearly two hours with Gillian and Sarah until it was time for dinner, and after that they had all gone out to the Union Bop, a mixture of dance and nightclub held in the Union each Friday night. Her friends had been predictably excited and envious and eager to advise her to see David again as soon as could be arranged. Next Friday was far too long to wait – they were confident that a chance meeting before then could be easily engineered.

Lucy had made it clear that she wasn't ready for anything serious so soon, although she stopped short of explaining that she didn't want to fall behind in her lectures so early in the semester. Gillian and Sarah were both taking courses in the humanities, in which, Lucy suspected, one could miss the odd lecture and still be able to understand what came later in the syllabus. In contrast, science courses, and particularly physics, were structured in logical sequences of interdependent ideas. Each night, she intended to work through the day's lectures so that the following morning's lectures might be comprehensible. To do otherwise, Lucy reasoned, would be like trying to construct a building without first laying the foundations.

This opinion, of course, she kept to herself. They wouldn't understand, or, worse, would think that she was claiming her precious physics was more difficult than their own subjects, which she certainly didn't intend. So she had gone along to the Union Bop cheerfully enough, grateful, after all, to have friends to confide in (if not completely), and, in fact, she had enjoyed the break. Through much of the weekend, though, Lucy had spent reading around the week's work, although, she had to admit, she had not encountered any difficulty with the lectures up to now.

So, just before nine o'clock on Monday morning, she entered the Mathematics Institute satisfied with her weekend's work, prepared for the new week's lectures. She was not in any way going to seek out David. Nevertheless, she found herself looking across to the adjacent Physics Building just in case he happened to be going in. As she walked towards the maths lecture theatre she felt a tap on her shoulder and she spun round. She felt a momentary pang of disappointment on seeing that it was George, the class spokesman, which must have showed, because he immediately apologised: 'Sorry, were you expecting someone else?'

'No. Yes. I thought it might be my friend, Gillian,' she lied. 'How was your weekend?'

'More to the point, how was yours?' he asked. 'Is everything OK after Friday?'

Lucy was surprised to realise that she had momentarily forgotten about the choking episode. It seemed to her that so much else had happened, that she had put it to the back of her mind. 'Oh, fully recovered. Thank you for asking.'

'Glad to hear that. Going to maths?'

'Yeah, I was just on my way.'

'I'll sit beside you, then. You never know when you might need help!'

'Huh!' Lucy was not best pleased at the attempted humour at her expense, but she could think of no way of demurring politely and so she led the way into the rear of the lecture hall with George following closely behind her. She headed for the first row they came to, at the back of the hall, but George suggested firmly that they go all the way down the steps to the front, and so she went after him, annoyed with herself for following him so meekly.

Her resentment was further provoked by his apparent inability to remain silent during the lecture. At first, she nodded her head at his whispered comments just sufficiently to acknowledge him and not be impolite, but, she hoped, not enough to encourage further low-decibel observations. It soon became apparent, however, that he was a *sotto voce* serial monologist and she thought to herself how apt was the *spokesman* epithet.

At the end of the lecture, she stood up quickly without acknowledging George and headed for the exit at the front of the lecture room in the corner, directly across from her. She walked rapidly down the corridor which emerged into the entrance hall already crowded with students transiting to their next lectures. She was due in the Physics Building in five minutes, but she realised that George, whom she knew was just behind her, would also be bound in that direction.

'Lucy Darling?' The man was standing tall, like a stepping stone in a river, unperturbed by the flow of bodies around him. He wore sliver-rimmed spectacles and was prematurely balding.

Instantly anxious but thankful for the excuse she nodded and turned to George. 'You go on. I'll see you later.'

'I'm Paul Evans,' said the man, as George was carried past them by the current. 'I'm a student, but I'm also a *Stat* hack.' Although the student newspaper appeared only fortnightly, Lucy was aware of it from Freshers' Week. 'I gather you had a brush with death on Friday?' he asked.

'What? Oh, I see. Well, I suppose you could put it that way…'

'One of the students at the tutorial contacted me on Friday with the news, but it was too late to come and find you then. She told me basically what happened, and she gave me your description – that's how I found you just now.' *Goodness – I wonder what she said!* He leaned forward so he could make himself heard against the background noise. 'Look, do you have maybe half an hour so I could write your story?' he asked, his eyebrows arched high in appeal, pushing his skin into corrugations that invaded the territory abandoned by his hair.

'No, sorry, I can't – I'm on my way to a physics lecture. In fact, look, I'm sorry, but I don't think I really want to be interviewed about this anyway.' She thought for a moment. 'Why don't you ask David Lane – he's the one who rescued me? He's a postgrad in this building.'

Paul splayed his hands in a Gallic shrug of defeat. 'Yes, I know. I already asked him this morning, but he didn't want to do it. I think he was just too modest. He suggested I ask you, which I was going to do anyway. You know, it was a wonderful thing he did. It would be a shame if it went unrecognised by the wider public.'

Lucy reconsidered. If she declined, then David would know that she had done so, because he had sent Paul in her direction. In that case, it might seem ungrateful of her not to have given him his due mention in the newspaper. She relented: 'OK, yes, you're right – what David did does deserve to be acknowledged.' She checked her watch. 'My morning lectures finish at one. I could give you half an hour then. Any later and they'll have closed the dining room. Where would you like to do the interview?'

'Meet me at the reception desk in the Computer Science Building?' As he departed, he squeezed her arm gently: 'You know, you're doing the right thing by David.'

When Lucy entered the physics lecture room, she was pleased to see that George was already sandwiched by students on either side of his seat, and she found herself a place several rows back from him. Pity his neighbours! She made a mental note to see if these same students ever sat beside him again.

*

36

At five to one, she made her way over to the Computer Science Building. Almost all of the buildings in the university's science faculty were on the same site, and it took her only a couple of minutes to walk to Computing from the Physics Building. Sure enough, Paul was in the foyer, waiting for her.

He came up to her. 'Lucy, thanks for coming. Would you like to come round to my office?'

Lucy followed him through a large, open-plan area populated by serried ranks of students, each communing with one of the computers on the parade of benches that spanned the room in parallel rows. The area was surprisingly quiet, given the large number of students: many seemed to Lucy to be almost in a trance, unblinking, hypnotised by the screens. They left the room at its far end and she followed Paul down a network of corridors to his office.

Lucy's first impression was that it must have been ransacked in his absence. The extensive desk was covered with books on computer languages, photocopied journal papers, cardboard files, computer motherboards, USB adaptors, cables and computer printouts. The computer itself emerged from this sea of clutter like an iceberg, its keyboard nearby, floating unevenly on the surface of all the mess.

In a worse state, though, was the floor, which was almost completely hidden by sedimentary layers of books and papers strewn across the room. Paul stooped to sweep a path to one of two chairs so Lucy could sit down. He smiled apologetically: 'Some people think I'm untidy,' he acknowledged, reaching over to switch on a coffee maker.

'Surely not!'

He laughed. 'But there is order amongst all the chaos. I know where everything is. I leave all this stuff around the place because when I need it, it's quicker just to pick it up than storing it all away neatly in filing cabinets. I'd be

37

wasting time if put everything back once I've used it because I just might need it as soon as I put it away.' He pursed his lips and looked doubtfully at the floor for a moment. 'Although, if I'm honest, sometimes I find papers buried in the deeper layers that I haven't touched in years.'

'You're lucky to have an office to yourself,' said Lucy.

'Yes, I suppose I am. All the postgrads in computing have one. I'm doubly lucky, of course, because this doubles as an office for *Stat*.'

'Oh, I'm surprised you don't have an official headquarters, or that sort of thing.'

'Ah, we do. We rent an office in the Students' Union. All the money comes from advertising – that's why the paper's free. But several of us who have our own offices also use them for the newspaper. That way, there's backup, redundancy if you like, the same as distributing your crucial servers widely on a network, just in case they close down the headquarters.'

'What, could they do that?'

'Oh, they could, and they have, in the very recent past.'

'Why, what happened?' asked Lucy.

'We were expelled from the premises for a week because of something we published,' Paul explained. 'Actually, it was my story – I was really proud of it. Still am, in a way.'

'What did you write?'

'I exposed a weakness in the University computer systems and network.'

'Surely that would make the University grateful to you! So why did they close you down?'

'Ah, well, they objected to the methods I used to get the story. You see, I hacked into emails and other systems that they thought were very well protected. I didn't do anything malicious, you understand, and I gave the details in confidence to the IT department.'

'I still don't see—'

'My undoing was referring to a private email from the Vice-Chancellor about tuition fees.'

'Oh, then I'm not surprised you were closed down. You were lucky not to be expelled.'

Paul pointed a finger at her to emphasise the point. 'Precisely. It was seen as a severe invasion of privacy. Anyway, to cut a long story short, the student population protested strongly over our shut-down and we resolved it out-of-court, as it were. But I'm still on unofficial probation!'

He reached under the desk, took out two mugs from the cupboard underneath and began pouring coffee into each. He handed her one of the mugs and a milk sachet. 'So what was it exactly that caused you to choke in the first place?'

Lucy described the whole episode as well as she could remember, with some prompting from Paul. She was impressed with how well he was able to extract details that had escaped her notice until now – for instance, the fact that nobody in the room had had the presence of mind to get out of their seat before David had caught her at the front of the class. Paul ended the session by asking if he could take a photograph, to which she agreed, with some misgivings.

'You won't put a spin on this, will you?' she asked.

'No, I promise you. Look, this story speaks for itself, straight down the middle.' He gave a lopsided smile: 'There's nothing political here. It might even go some way towards my rehabilitation with the University Court!'

The interview over, she asked him on her way out of the building when it would be published.

'Look out for it on Friday,' he replied. 'I owe you one. Thank you, Lucy.'

*

With each day that passed that week, Lucy's hopes of bumping into David diminished, but the upside was that

Friday was drawing ever closer, when she knew she was bound to see him in the tutorial and, she hoped, maybe afterwards as well. In any case, thinking about all the new material in the lectures and reading around it and exploring it seemed to make the days pass more quickly.

On Friday morning as she headed for breakfast in the hall dining room she took a free copy of *Stat* from the bundle in the foyer. It was with a sense of unreality that she studied the large photograph of herself smiling out of the front page under the banner headline: *Escape from Death in Tutorial*. She read the caption under the photograph: *Horrified students in a physics tutorial watched as Lucy Darling (above) nearly choked to death on a gobstopper. Physics postgrad tutor David Lane saved her life using abdominal compressions.* The accompanying article recounted the events as she and the other student had told them to Paul. He hadn't needed to sensationalise the story – it was as accurate as she recalled it herself, and, she had to admit, it was arresting enough without requiring any embellishment.

When she entered the dining room several students whose faces she knew only slightly acknowledged her like an old friend: each had a copy of *Stat*. She thought this must be what it would be like to be a celebrity – you keep meeting people who think they know you, but actually they only know your picture. Then it registered that at least celebrities might claim to have earned their fame: she, on the other hand, had been an entirely passive party in the story.

By the time she approached the Mathematics Institute for the first lecture of the day, she was feeling distinctly uneasy about facing her classmates – would she have to run the gauntlet of her paper-waving peers to reach the lecture room? Sure enough, she was met by several knots of excited students, each grouped around a copy of *Stat*. Many of these she knew from the tutorial group, and the others she

40

recognised as taking maths but not physics. Their comments were kindly meant, overall, but Lucy had already been the butt of enough mild teasing on Monday to feel only embarrassment now, and a wish to put the whole experience behind her.

The one positive outcome was that she was able to use the retinue that she had attracted to shield her from the attentions of George. Several times during the week he had attempted to repeat Monday's manoeuvre of sitting beside her, each time without success. Now, she thought, if she could just keep the advantage until the day was over, he might get the message and stop pestering her next week.

The tutorial started at two o'clock but Lucy decided to hang back until the last minute so that she would have the excuse of the filled front seats to go to the back of the room as she had done the previous week. Too late, she remembered that, because you entered this room from the front, she would be seen by everyone as she came in. Sure enough, as she walked through the doorway, all eyes were on her and to her horror, they started to cheer her. She forced a smile and waved vaguely at the crowd as she made her way up the aisle to the back row.

Just as she sat down in the very same seat as last week, she was startled by the sound of someone settling into the seat next to her. George had apparently left his seat in the front to follow her all the way.

'I couldn't let you sit here all on your own,' he grinned. 'You might get flashbacks.'

She was thinking frantically how she might politely ask him to go back to his own seat when David came in and acknowledged the class.

'He's probably been on a first-aid course since last week,' George whispered to her. Lucy ignored him, furious that he had done this.

'Brought any sweets today?'

She whipped round to face him. 'No,' she hissed. 'Now would you mind being quiet, please? I want to hear what he's saying.'

She was immediately embarrassed by what she thought was probably an over-reaction, but at least it seemed to work because George managed to remain quiet for all of half an hour before resuming his whispered comments about the tutorial.

Lucy tried her best to disregard him and kept hoping to catch David's eye. As soon as he would look at her, she was ready to flash him a smile. She thought about raising her hand to take part in the question-and-answer session, but, somehow, the very presence of George next to her had destroyed her concentration so that she just had to sit there, quietly fuming.

As the tutorial drew to a close, the constant attentions from George had pushed her to the limit of her patience. David was describing the four forces of nature: '—which leaves the strong force, key to the fission process in nuclear power stations for example.'

George leaned over to Lucy and whispered 'Actually, it's the good old electrostatic force that releases the energy in fission – the strong force is really trying to keep the nuclei from splitting.'

Lucy rounded on him. 'George,' she whispered angrily, 'I thought I made it clear that I wanted to hear him.'

She turned her attention back to David and was appalled to see that he had caught the two of them whispering. So dismayed was she that she couldn't bring herself to smile back at him as she had planned. From David's point of view, all that he had seen was two students sitting at the back, apart from the rest of the class, whispering together and not paying attention to his tutorial! She was too disconsolate even to feel anger at George any more.

The tutorial ended soon after. David thanked the class, picked up his papers from the desk and left the room, which

42

gradually emptied after him. As they stood up together, George said, 'Look, I'm sorry if I distracted you today.'

He was repentant, then! He had come to his senses at last. For a second, Lucy was about to reply, forgiving him, but he continued: 'I hadn't realised you had such a thing going for him.'

At this, Lucy's anger nearly erupted, but she controlled it, saying nothing, merely gesturing sharply for George to move on ahead of her.

She left the Physics Building and carried on walking without any particular destination. She wandered away from the town along the paths that dissected the wide open fields of the science campus. She gazed westward unblinkingly into the vast, pale-yellow sky of the October afternoon. She hadn't realised just how much she had been looking forward to David's company until the chance had been taken away from her. And, yes, she had to confess, it was his company and not just the opportunity to talk about physics that she had been hoping for. But now he would assume that she was just the same as any other eighteen-year-old girl, whispering and flirting with the boys in her class, not paying attention while he tried to teach the subject closest to his heart.

Her eyes welled up in a sudden wave of wretchedness. *Oh Dad*, she thought, *if only you could cuddle me now and make it better* – and then her sky fragmented into a kaleidoscope of yellow lights.

CHAPTER 4

Friday 8 October – Monday 11 October

David had collected a copy of *Stat* on his way into the Physics Building that morning and taken it straight upstairs to the office that he shared with Mike. Normally, neither he nor Mike would have arrived before ten o'clock, but David had been coming in a whole hour earlier every day that week, and so he had the office to himself. He settled down in his chair and unfolded the newspaper. Sure enough, Lucy's steady gaze met his as he studied her photograph intently.

He read neither the caption nor the story; he simply stared, mesmerised, at Lucy's face. He had thought he had already captured her image in his mind's eye in exquisite detail, but the picture seemed to add a dimension to his memory of their brief encounter, transporting him back to relive the sweet hour that he had shared with her.

David had noticed her as soon as he had walked into the room to give his first tutorial of the session a week ago. She was sitting alone in the back row, the isolation itself adding to the impact she made on David. He was immediately captivated by her long, straight chestnut-brown hair, and, even from the back of the room, her eyes seemed to be looking deeply into his own, even into his mind. He had spent much of the tutorial trying not to stare at her, and so he had risked only momentary glances, blinking after each surreptitious glimpse like a camera taking a photograph, storing her image on his retina and stopping others from overwriting it.

As the tutorial had progressed, so did his disappointment grow that she wasn't taking part. His delight when she eventually raised her hand was transformed into near incredulity at the profundity of the question she asked. Taken aback, he forgot even to ask the class to suggest an answer. Indeed, for a second, he wasn't certain himself how to proceed until he had turned away from the class and begun to scribble equations on the whiteboard.

The moment he derived the answer was revelatory – that the electrons would have to travel at the speed of light attested to a connection with relativity which was familiar to him but which, for some reason, he had never applied explicitly to such a simple scenario as two electrons travelling side by side.

His recollection of the next few minutes was disjointed: instead of replaying it in a logical stream in his mind, he could only conjure up a jumble of frozen images, like spotlit *tableaux vivants* in an experiential museum. He *did* recall enfolding her in his arms from behind, and thrusting with his fists upward into her diaphragm, all thoughts of her beauty obliterated by the utmost urgency of keeping her alive.

His suggestion of going for coffee had been prompted only by his concern that there might be after-effects either to her throat from the errant sweet or, indeed, to her abdomen or ribs from the force he had used. It was certainly not because of her undeniable beauty, or even the insight she had demonstrated in asking her question. In fact, if she had not choked, the last thing David would have considered would have been to ask her out.

The trouble was that David was generally shy in female company, sometimes painfully so. The issue had dogged him throughout his school days and into adult life. True, he had had no problem with speaking to the girls in his class at school, but they were like family, surrogate sisters he had grown up with over six years of secondary education.

The only release from his introversion was in his dreams at night. Often, he would be wandering through some vaguely recognised town, not exactly the one nearby his home, but with some of its components transplanted, like the school or the hill to the east, and he would meet a girl, lost, like him. In his dreams, her features were never clear. They would strike up a conversation, and discover that they shared a remarkable number of interests and had the same yearnings in common. The dreams would culminate in an achingly tender kiss, and when he awoke, he would be convinced in his sleep-befuddled mind that they had really met. The personality of the girl was so real to him that, despite his scientific rationality, he could believe that she really existed, and was searching for him as he was for her.

In the late summer evenings before he went up to university, he would climb to the top of the hill overlooking Fort William and its satellite villages strung out along the margins of Loch Linnhe, and sit watching the sky to the north-west reflect the gradually shifting spectrum of the setting sun. As the early stars twinkled into existence from out of the deepening violet, the lights of the darkening town switching on in counterpoint, he would suspend logic for as long as he could, and persuade himself that his dream-girl was searching for him here on the hill. At last, he would sigh heavily and rise to his feet, and begin the long trek eastwards down towards his home in the dark, turning his back on the town, now a burning galaxy that outshone the others overhead.

His only solace as a teenager was to escape into physics. Science fiction was an aperitif that stimulated a hunger for the hard science – usually physics – underlying the stories, and he found that he had a natural inclination for the subject. When he first came across the formula for the energy density of an electric field – never even mentioned in his school syllabus – he devised a spaceship where switching on an electric field generated an energy density, and, hence, a gravitational-mass density, that would warp space sufficiently to be used as the *hyperdrive* of his

46

science-fiction books. He realised, of course, that the effect would be miniscule in reality, but this led him to question just to what extent space could be distorted, and so he began gradually to teach himself the difficult subject of general relativity.

There were few girls in his class at university, and the number diminished as he progressed through the years. His first year of research had consumed all of his time, and so here he was, nearly a quarter of a century old without ever having had a proper girlfriend! So he was enchanted to discover that Lucy had read science fiction when she was young, just as he had. Moreover, they both shared a passion for relativity and even seemed to view the world in the same way. Their only difference appeared to be her belief, and his non-belief, in a deity, but even that subject they had debated with mutual respect. She was the first person to whom David had dared even to hint about his doing work in parallel with his PhD. She was clearly intrigued, and, to his joy, she seemed to welcome the prospect of a meeting again after the following week's tutorial.

However, over the weekend, his raw elation had gradually scarred over into a familiar crust of self-doubt. How could he have been so deluded as to think she could be interested in him? She could only be eighteen, nineteen at most. To her he must look like an old man! He had never in his life been so close to such an attractive girl: she was bound to have a boyfriend, or, if not, she could have her pick from the young men of her own age who would inevitably swarm towards her like bees round the honey-pot.

He had confused the interest she had shown in discussing relativity with an interest in his own company. Of course, he could see it now – she had admitted it herself – she found it liberating to talk to someone about her passion for the subject and find out more about it. David was simply the opportunity – the instrument – for providing that instruction.

Despite himself, David could not stop thinking of Lucy throughout the whole weekend. It was as though proximity to such a perfect creature had burned an after-image into his brain, disturbing his neurons so that they continued to fire long after the stimulant had gone.

On Monday morning, he awoke uncharacteristically early and set off for the Physics Building before nine o'clock. As he approached the science campus, it dawned on him that Lucy would have maths for the first lecture of the day, as all first-year maths and physics students did. So he gradually slowed his pace until he was at the top of the road overlooking the campus, from where he could see the Physics Building, the Mathematics Institute and the path to her hall. There was a bench at the side of the road and he sat down on it, feeling like the old man she must think he was, but it would be less conspicuous than standing.

Sure enough, a few minutes before the lecture was due to start, he could see Lucy coming along the path from the hall, heading for the Mathematics Institute. He could only make out her long chestnut hair at this distance, but his memory filled in the details, and he watched her like a love-sick teenager all the way until she was swallowed up in the entrance to the department. Wearily, he rose and walked down to the Physics Building to begin his day.

This pattern was repeated every day that week, culminating in picking up a copy of *Stat* on Friday. He was slightly more sanguine by that time, and looking forward to seeing her in the tutorial. He had long since convinced himself that she had understood his parting words – 'see you next Friday' – to mean simply that he would see her in the tutorial, that it was not a proposal to meet afterwards. However, a corner of his mind still dared hope that he was wrong, and that she would wait behind, after the others had left the tutorial, and that they would resume their conversation, maybe even walking into town together.

David knew he had lost her right at the start of the tutorial. She was sitting in the same place as before, at the back of the room, but this time, she was not alone. The two of them sat there together, Lucy and one of the students whom he had noticed last week, quite an extrovert, he recalled. As in his first tutorial, he tried to avoid looking in Lucy's direction, but he could see them at the edge of his vision occasionally whispering to each other.

He gave what he knew was a fairly pedestrian tutorial. He couldn't concentrate on the topics and was unable to summon up the excitement of his performance of the previous week. The other part of his mind was preoccupied with a mixture of feelings. Uppermost were self-pity and a sense of betrayal. He was particularly stung by the fact that she should be heartless enough to allow herself to be distracted by her student friend while he was teaching. At the same time, he was feeling foolish for having allowed his imagination to wander in such fanciful directions. This was the penalty for dreaming – waking up to reality was all the more painful.

As he closed the tutorial, he thought for a moment about sticking to his original plan, waiting until the students had gone. Then he decided there was no point in inflicting further pain on himself. In a twisted way, he hoped that, by walking out first, he might induce a fleeting pang of remorse in Lucy, but then he realised that she would barely notice his departure.

*

At ten o'clock the following Monday morning, David and Mike were propped high on laboratory stools in the middle of the optics laboratory where Mike worked for his experimentally based PhD. David had reverted to his

49

normal waking time and had arrived in the building just a few minutes before. His pigeon-hole at the reception desk had contained a note circulated to all postgrads informing them that some of them would have to admit another postgrad student into their offices, because the department's research success had led to an increase in postgrad numbers this year. His first thought was to find Mike to see his reaction.

While Mike read the note, David admired the massive, highly polished, black granite slab which served as a vibration-free base for Mike's complex and delicate optical components. The slab had been one of a job lot acquired by a former head of the opto-electronics section from the same firm that supplied gravestones to churchyards in the vicinity, and represented a considerable saving on the cost of commercially available optical benches. Although he had seen it many times before, it occurred to David for the first time that the highly reflecting granite surface might not be the safest texture to place in the vicinity of the intensely powerful laser beams that would zig-zag their way through the optical components, and he was about to comment on this when a knock on the laboratory door caught their attention.

'Come in, it's OK,' shouted Mike above the constant whine of the vacuum pumps that were used to maintain the low gas pressures of his plasma cells. The door opened fractionally, and the two were surprised to see the head of Professor Jeremy Wilgoss insinuate itself through the opening and peer round the room like a nervous rabbit checking out the territory, before the professor appeared to summon courage and venture fully into the laboratory.

'Hi, Jeremy,' David greeted his supervisor. 'You lost down here?' he added, in ironic reference to the stratified order within the Physics Building: theoretical physics offices at the top of the building, experimental laboratories in the basement, and administration and most of the lecture rooms sandwiched between them.

Wilgoss gave a bleak smile and advanced into the room with his characteristic slow gait, straightening each leg in front of him as he walked, almost goose-stepping, towards David and Mike. He paced a complete circle around the two students on their stools, his eyes taking in the equipment and surveying the whole laboratory in his circuit. 'Just checking out where the Morlocks live,' he answered, coming to a stop in front of them.

'We're honoured by any interest from the Eloi,' said Mike, catching the reference.

'Hmm ... I must say, you do seem to live in a most hazardous environment,' said Wilgoss, scanning the room again. He read out the safety notices placed prominently around the laboratory. 'High voltage, danger of death; Class 4 laser radiation controlled area; Warning, ozone hazard; Gas cylinders, explosion hazard. So what's the danger with the gases – I thought you just used noble gases? They're not explosive.'

The notice was attached to a door at one side of the laboratory. 'No, it's not that,' said Mike, 'it's in case of a fire. The temperature in a fire can increase the pressure in a gas cylinder so much that they can explode. That's why we keep all our cylinders in the small room behind that blast-proof, fire-proof door out of harm's way. You're right about the gases: they're all inert.'

'So, in an ironical sense, the disaster would be self-limiting,' said Wilgoss. 'If the cylinder explodes, it will release inert gas which will promptly snuff out the fire!'

'Yes, I suppose it would,' Mike responded without enthusiasm. He probed cautiously for some indication of Wilgoss's reason for coming down. 'Would you like me to show you round the laboratory? Is that why you're down here?'

'No, actually, I've had an idea,' Wilgoss stated, leaving an expectant pause with the studied eccentricity that David recognised from the few encounters he had had with his supervisor. David thought facetiously that his statement ought to be marked by a celebration of some kind, as Wilgoss was not one to blaze a trail with his work.

To David's dismay a year earlier, he had been told that Wilgoss would supervise his PhD. Since David had been an undergraduate at the university, he already knew of Wilgoss's reputation as a hopeless postgraduate supervisor, but he had had no choice in the matter. Sure enough, over the past two years, Wilgoss had shown no real interest in David's project, and had let him get on with it largely unmonitored. This behaviour continued even after the disaster in spring, when one of Wilgoss's students had failed his PhD *viva* because of a fundamental error of physics.

The degree of PhD was awarded in recognition of original research work presented in the form of a thesis. The contents of the thesis were judged by two examiners who would study the work meticulously and then question the student in depth at an exhausting *viva* examination. It was practically unheard of for a student to fail at this stage, because a conscientious supervisor would have detected any major blunder in the student's work long before the preparation of the final draft of the thesis. Indeed, it was one of the supervisor's responsibilities to keep themselves informed of the progress of the research and to remain alert for mistakes that might have a bearing on the validity of the work. However, it appeared that Wilgoss either had not been alert enough to spot the fallacy or had not kept himself sufficiently up-to-date with his student's work to avert the disaster. Either way, it had put Wilgoss in a very bad light when his student had been offered a compensatory MSc, which was poor comfort indeed for the loss of a PhD and three years' wasted effort.

Mike was forced to fill the awkward gap. 'Is your idea something to do with optics? Lasers? You want to test out your idea?'

'Ahaa! We shall have to see!'

Embarrassed, partly because he had not, in over two years of postgraduate study, ever spoken to Wilgoss, knowing his reputation for awkward conversation only from his fellow students, Mike slid off the stool and prepared to show him around the laboratory.

Wilgoss showed particular interest in the large, spherical stainless-steel vessel that sat centre-stage in the room. It bristled with probes, monitors, data cables, power cables, gas hoses and glass tubes, all apparently focused towards the centre of the device. 'I take it this is the famous plasma vessel.' It came out as a statement rather than a question, as though he couldn't bring himself to admit not knowing something, but Mike took it at face value and answered that yes, it was the plasma cell that was key to his laser-wakefield-acceleration experiments.

'And these are your lasers,' Wilgoss added, pointing to several long metal boxes marked with laser sun-burst symbols. 'Are you using them all?'

'No, just those two,' said Mike, indicating two particularly large lasers close to the plasma vessel. 'That one creates the plasma wave, the wakefield, and that one then generates new ionization – new electrons – within it, so they can be accelerated. The others are just spare lasers, much less powerful, though.'

'Then we have a job for one of them,' announced Wilgoss. 'I've seen Tony,' he added, referring to Professor Tony Packard, the head of opto-electronics research and also Mike's supervisor, 'and he's happy for me to commandeer your services to push back the frontiers!'

'What, you mean Tony wants me to set up an experiment for you?'

'Indeed, that is exactly what I do mean.'

'Oh, I see. Well, yes, of course, that's fine. I mean, that's very good.' Mike paused. 'Um … will Tony let me know what to set up, exactly, or shall we be working together, you and I?' David could guess that the cause of Mike's discomfort was that he was uncertain of the protocol – shouldn't he wait for Tony's permission to take time out helping Wilgoss? Indeed, Mike would probably not particularly welcome an interruption to his own PhD work, especially as he had spent much of the last year exploring a blind alley and he badly needed to recoup the lost time.

'Your loyalty to your supervisor is commendable, and, of course, you may speak to him about it. In the meantime,' said Wilgoss, handing Mike a folder, 'I have prepared an outline of the experiment that I should like you to set up for me with your laser.'

Mike opened the folder and took out the drawings. 'So, can I ask what the experiment is meant to do?'

'Yes, you may well ask. Actually, I believe that it may settle a hundred-year-old question about the Abraham-Minkowski stress tensor. To put it simply for you as an experimental physicist' (Wilgoss emphasised the word *experimental*) 'it has never been clear whether the momentum of light photons in transparent materials increases or decreases compared with their momentum in a vacuum. Extraordinary, I know, but that's the fact of the matter. I hope that my experiment may resolve the issue.'

David was tempted to say something to the effect that, if the theoreticians had been arguing the case for a century, then it was one up for the experimentalists if they were able to decide the issue. However, he held his peace, knowing that the humourless Wilgoss would not appreciate repartee that he hadn't initiated.

'Let me know if there is anything you do not understand in the drawings,' said Wilgoss and then, almost robot-like, he turned on his heel and walked slowly towards the door. 'I shall return in a fortnight's time,' he proclaimed without turning round, his voice raised against the background noise, and with that, he left the laboratory.

'How do you put up with him?' asked Mike.

David shrugged. 'We see each other as little as possible, which probably suits us both equally. You know, he may not put your name on the paper, if there is one. He may not even mention you in the acknowledgements.'

'Oh, that's OK. To be honest, I'd really rather not be getting involved with this at all. I'll see what Tony has to say about it.' Mike looked at his watch. 'Come on, it's coffee time,' he said, and they both left the laboratory for the coffee lounge.

*

The coffee lounge was situated on the ground floor, one floor above the level of the laboratories. Congregating within the coffee lounge was much more than a simple act of queuing for a maintenance dose of caffeine: it was the gathering together of practitioners of what is often a solitary, reflective profession – and included both experimental and theoretical physicists. Not only did this punctuation to the morning provide an opportunity to talk with friends and colleagues, but, at its best, the assemblage could create a critical mass of minds capable of sparking new ideas off each other or reining back the more outlandish ones. In the coffee lounge, physicists could learn of a development in another field that might solve a problem in their own. The simple change of focus from equations on the notepad or equipment on the laboratory bench to a wider perspective for half an hour was a brain tonic, and physicists returned to their work after coffee refreshed, their minds more open to speculative ideas.

The coffee lounge was supposed to be egalitarian, used by lecturers, readers and professors as well as the lowly postgraduate students. Nevertheless, there was, by common consent, a hierarchy in the arrangement of coffee tables. The current head of the physics department, Professor John Oakhill, always held court at the table closest to the large window that afforded what once had been a panoramic view of the campus, now completely obscured by the new medical-school building. He was regularly joined around the circular table by the senior members of the department – the other professors, readers and senior lecturers – and conversation would generally be university politics, local and national, or, for light relief, gossip from the campus and from departments in rival universities.

To withdraw from the table at the window and walk back into the lounge was to descend the departmental ladder, past several tables of young lecturers discussing timetables, curricula and grant applications in equal measure, beyond the table of post-doctoral fellows preoccupied with time-limited projects and their next fellowship or maybe a lectureship, to reach the postgraduate students at the table furthest away from Professor Oakhill's. Of all those frequenting the coffee lounge, the postgraduate students were, perhaps, the most free to think about physics with thoughts unsullied yet by academic jealousies and largely untroubled by deadlines except towards the end of their three-year studentships.

Mike and David filled their cups from the urn and sat down at the postgraduate table, which was already fairly crowded. At a lull in the conversation, Mike announced that he had been visited by Wilgoss.

One of Wilgoss's other students was curious. 'What was he after? Did he want you to throw some light on his problems?'

Mike smiled. 'Well, actually, yes, very literally, as it happens. He wants me to set up an experiment with a laser. Apparently there's an unresolved problem a hundred years old, and it could be decided by an optical experiment.'

56

Mike proceeded to explain the little he had learned from Wilgoss about the problem. 'He's left me a folder with notes in it, but I couldn't face studying it without a jolt of caffeine.'

'You can bet that whatever is in the notes, you'll need to put a lot of your own time into it to get it right,' said Roger, a solid-state-physics postgraduate student in his third year. 'What did Tony Packard say?'

Mike looked despondent. 'I haven't actually spoken to him yet, but Wilgoss has, and seemingly he's in favour of my helping out. I'll check with him, of course, but I don't think Wilgoss was making it up. Anyway, I've really got no choice – I have to do this.'

'Well, of course you have a choice,' said Roger. 'In fact, it may turn out to be a bit of good luck for you, if it goes well, but you have to balance that chance with the time it would take you.'

Christos, a mature student in his final year as a postgraduate student in theoretical physics, spoke for the first time. His English was pedantically slow and precise, and he had a habit of inserting his sentences into the conversation in the manner of a tank pushing through a crowd: the crowd would part before him and he would speak into the expectant silence. Since these interjections were often rich with new perspective on a problem, they were tolerated and sometimes welcomed by his audience.

'You see, of course, that Mike has no choice. Mike has not free will – I have not free will. In Mike's brain, there are now several local networks of neurons, each of which carries the summation of a different aspect of the decision that Mike has to make. One of these aspects is the possibility of scientific recognition for his part in solving a famous problem. Another aspect is pressure from Wilgoss and maybe also Packard. There will be a large disturbance in the local neural network that represents Mike's concern about the time this will take him. If we could look into

57

Mike's brain and see the different states of these local networks, we could predict the outcome of the combination of all of the states, which is the decision that Mike will make. Therefore, if you define free will as the ability to make a decision that could not have been predicted even by an imaginary omniscient observer, then I believe that there can be no free will. This is because, in principle, an omniscient observer can see the state of the neural networks and correctly predict the decision that Mike will make, even although Mike himself may not be conscious of the mechanism behind the decision process.'

Mike took up the challenge. 'OK, Christos, say you can look into my mind and you see how the potentials from the various local networks, as you put it, are shaping up. Say you reckon I'll decide to go along with Wilgoss. I might decide not to, just to prove you wrong! So, in the end, I have free will, because I have the last word.'

Christos's swarthy, sun-weathered face broke into a broad smile, the crinkling glee radiating from his eyes with the joy of debate. 'My friend, you cannot escape in this way. If you ask me, I, who can look now into your mind and tell you what you will do, then this knowledge of my prediction will mobilise another local neural network in your brain, which contributes to the decision you make. As the observer, I now see that this extra potential is enough to change your mind, and that you will change your decision to spite me!'

'If Mike doesn't have a choice,' said Roger, 'and it's all predictable, pre-determined, then if Mike commits murder, there's no point in punishing him, is there, because he had no choice in the matter?'

Bill, the theoretical postgrad, joined in: 'Ah, that takes us into an entirely different debate. That begs the question of why we punish. If it's to restrain the murderer from doing it again, or to make relatives of the victim feel that just retribution has been meted out, or if it's to encourage others not to murder people, then I guess there might indeed be some point in punishing a murderer. In fact, now that I think of it, punishment for the murderer will also re-set the action thresholds for his decision-making process, so everybody wins!'

'Except the victim!' said David. 'But say Christos is omniscient – actually we know already that he's omniscient,' he continued, and Christos's smile broadened, '– then does Christos himself have free will? No, wait a minute, suppose that a god exists who is omniscient – does that god have free will?'

Christos began ponderously. 'I would have to say that the answer has to be no. But my reason is not what you might expect. My answer is no, because there can be no such thing as a truly omniscient being in the god-sense that you mean.'

He held up his hand to acknowledge and quell the outcries from several seats around the table. 'Let me explain. If such a being could know everything, this must also include knowledge about itself. But to know a thing is to have a model of the thing in the mind. There has to be a model so that different scenarios may be acted out in the mind using the model before these scenarios are implemented in reality. An omniscient being must have a model of the whole universe in its mind. But remember that the universe also includes the being itself. So part of the being's mind must contain a model of itself. But if the model is accurate, the model will include *not only the being's mind but also the model inside the being's mind.*' Chrisos's hand gently beat out the metre in the air for emphasis, palm upward, middle finger and thumb touching. 'So now we have a model within a model. But inside that model must be another model and so on like Russian dolls to infinity.'

'Christos,' exclaimed David, 'what about this for an illustration – think of your omniscient being as a TV camera connected to a TV set that shows what the camera sees. Then point the camera at the TV set and the screen will show the room with the TV on a table in the room. But the picture on this TV set on the table will be the room with the TV set on the table. And the picture on this TV set will include the TV set – and so on, to infinity!' He recalled something Christos had said about acting out different scenarios using the model. 'Oh, and so that the being can manipulate the model in its mind, the TV set pictured on the screen could show a possible change to the environment – maybe the table collapsed under the weight of the TV which is now shown tumbling onto the floor.'

'Ah, David, you distil the essence of my argument very clearly,' complimented Christos, beaming. David could not help being flattered by the praise. The Greek was ten years older than most of the postgraduate students, and his relaxed air of worldly wisdom was an accretion of his days of national service in the Greek military and subsequent voluntary service in a hospital in Malawi before deciding that his future lay neither in fighting nor in healing but in the cerebral vocation of theoretical physics.

'So, using David's most vivid illustration,' he continued, 'we see that the true model of this being can never be constructed in its mind, because you never get to the final television set. There will always be part of the universe that contains the model of the universe which still needs updating. It was for this reason that I suggested that no being could be omniscient in this universe.'

There was a pause while everyone digested this. Then David spoke: 'Yes, I think I see what you're getting at. This idea of the being's mind containing a model of itself is a definition of self-awareness. In fact,' he said, becoming animated, 'because you're talking of a model within a model and so on to infinity, you're defining higher and higher levels of consciousness!'

'Yes, this may be true,' replied Christos, 'and you might then say that God would have an infinite level of consciousness,' he went on, playfully. 'But I am saying that these very infinities would rule out an omniscient god.'

'Not necessarily,' said David. 'If you accept that the model needs to contain only the essential elements of the mind, then you only need a fraction – say a quarter of the mind devoted to the model of itself. Then even an infinite series would only occupy – what – only a third of the mind.'

Christos's grin broadened. 'There you have the problem! Because you can only have the infinite series if the model is not full, not perfect. But then you have an imperfect god!'

This provoked further responses from the group about the definition of a god, and David listened with half an ear, but his mind was preoccupied with the coincidence of how closely this discussion came to what he had been working on since learning about Gödel's Incompleteness Theorem six months before. In fact, he was sure that a by-product of his work would show Christos to be right, but for a more subtle reason than Christos's infinite Russian-doll illustration.

However, he had decided to keep quiet about his project, for two reasons. If there were a flaw in his reasoning, he didn't want the humiliation of being exposed in front of so many of his peers. Alternatively, if he were right, he didn't want anyone else working on the problem and maybe solving it before he could. For a moment, he wished he could share his thoughts with Lucy, even just to get them straight in his own mind, and then he shook himself mentally, and tried to put Lucy to the back of his mind.

CHAPTER 5

Monday 11 October

The inaugural lecture later that Monday had been widely publicised, and was scheduled for the late afternoon, so that both staff and students could attend. The Physics Main Lecture Theatre was one of the largest on the whole campus. By five o'clock, the vast amphitheatre was nearly full, with undergraduates occupying more than half of the hall, the remainder being taken up by lecturers and postgraduate students, many of them from outside the Physics Department. From her seat half-way up the left side of the gallery, Lucy studied the audience. The middle of the front row had been claimed by a coven of the university's *éminences grises*, their black carapace gowns and hoods lined with vibrantly coloured silk suggesting a line of exotic, academic beetles.

At the very moment that Lucy spotted David Lane sitting half-way up the opposite wing of the auditorium, the background babble of the audience died to an attentive hush as the Vice-Chancellor walked across the stage to the rostrum. He wore a peacock-blue gown trimmed with gold; a black, velvet, John-Knox trencher with a silver tassel capped his fine, white hair. His gait was slow and measured as though labouring under the weight of all his academic regalia.

The Vice-Chancellor reached the rostrum, placed his hands lightly on its sides and raised his head high to survey the lecture hall, his eyes scanning the assembled crowd in a slow, lighthouse-beam sweep of the amphitheatre.

'The inaugural lecture of a new professor has a long and distinguished history at this university,' he began. 'The inauguration of Professor Jeremy Wilgoss is entirely in keeping with that tradition.' He paused, and looked over to Wilgoss, who was sitting on one of two chairs that had been placed at the side of the stage. 'Another aspect of the inaugural lecture at this university is the tendency for new professors to defer delivering their lecture for many months, or even years, after their appointment to their chair. I am pleased to say that, while Professor Wilgoss has managed to uphold this particular tradition, he does not match the record, held since 1978 by a professor in another faculty, who deferred his lecture for so long that he had to combine it with his valedictory speech on the day of his retirement.'

The Vice-Chancellor paused to wait for the good-natured laughter to subside, and then continued to introduce Wilgoss. 'I am delighted now to ask Professor Wilgoss to deliver his inaugural lecture, which he has entitled *Relativity in a bucket – the origin of centrifugal force.*' Wilgoss then rose and walked over to the podium, taking the Vice-Chancellor's place.

'Thank you, Vice-Chancellor. It was, indeed, a temptation to attempt to postpone this lecture for as long as I could get away with it, but the subject continues to expand, so that I feared that, by my retirement date, I would be forced to deliver not one, but a whole series of inaugural lectures, which would rather defeat the object of further procrastination.'

He paused to sip from a glass of water that he produced from a shelf in the rostrum. 'It may come as a surprise to some in the audience that a lecture dealing with the origin of centrifugal force could have been topical even last century, let alone today, but I propose to take you on a journey of the mind that begins in the 17th century with a bucket of water set spinning by Sir Isaac Newton, right up to present-day relativistic views on the matter, with a nod to projects considered worthy even of the National Aeronautics and Space Administration.'

At the mention of relativity, Lucy's attention, which had strayed from the rostrum towards the far wing of the auditorium, swung back again to focus on Professor Wilgoss. She could see no other students from her class around her, and she suspected that the majority of undergraduates in the audience were from the third and fourth years, but she herself had, of course, been attracted by the intriguing title referring to relativity, which was why she had made such an early effort to find out about the time and venue, nearly being late for the first physics tutorial as a result.

As Wilgoss's style changed up a gear from its slow pomposity, and he began to warm to his subject, she began to focus on what he was saying. As she understood it, the conundrum posed by Newton was essentially to ask what the difference was between water rotating in a spinning bucket and the water remaining still in the bucket but the rest of the universe rotating round the bucket of water instead. Since the two scenarios had the same geometry, did this mean that a rotating universe would pull the water to the sides of the bucket, mimicking the centrifugal force that is normally cited as the cause of the water surging up the sides of a rotating bucket?

Wilgoss had gone on to describe the contributions of physicists to this problem since Newton had formulated it – Mach, Einstein, Thirring and others, and then he had described tests of proposed solutions which NASA had been conducting using exquisitely sensitive observations of its satellites around Earth. Apparently, Wilgoss himself had contributed to the design of the latest test that NASA intended to launch in orbit around Jupiter.

Lucy was enthralled. Here was a very basic question, its simplicity highlighted by the very ordinariness of a bucket of water – she could picture quite vividly the ancient wooden-slatted bucket held together by rusty iron hoops. As Wilgoss wound up his lecture, she imagined the water rotating, contained by the steep sides of the bucket preventing the water from spraying out in all directions. The water at the sides of the bucket was continually being accelerated back towards the centre of the bucket but never reaching it— The enormity of an idea as it hit Lucy almost made her squeal out in the vast auditorium – surely she must be wrong! She went over it in her mind – no, she was right! So why hadn't Wilgoss referred to it in his lecture?

Lucy was now so engrossed with her idea that she barely registered that Wilgoss had finished, and it was only the applause, which she joined in automatically, that switched her attention back to the ceremony. The Vice-Chancellor had joined Wilgoss at the rostrum and was thanking him and saying how he never realised that such an every-day phenomenon as a rotating bucket could hint at a question of literally cosmic proportions. 'It is not usual for those delivering inaugural lectures to invite questions,' he continued, 'but Professor Wilgoss has very graciously volunteered to answer any questions on this occasion.'

As Wilgoss and the Vice-Chancellor searched for a raised hand, Lucy's heart started to thud fast. Dare she ask her question? She was genuinely excited by what she thought she had deduced from the lecture, but what a public humiliation if she were wrong in front of such a large and distinguished crowd! On the other hand, if she were right, it would surely get her noticed by the physics lecturers in the audience.

No other hand had gone up yet, and the Vice-Chancellor had begun to turn towards Wilgoss. She took a chance! Her hand shot up, almost of its own volition. The Vice-Chancellor saw it and pointed directly at her. 'Yes, over there!'

Lucy didn't stand up, but spoke out loudly. 'Professor Wilgoss, do you think that the force needed to accelerate matter in a straight line has the same origin as centrifugal force? Can that be explained by the rest of the universe dragging on it in the same way that it does on a spinning object like the bucket of water?'

There! She had said it, and it had come out quite well, succinctly, and she hadn't stumbled over any of her words. Wilgoss, who had not seen her raised hand, had spent the first part of Lucy's question trying to locate the source of the voice. Having done so, he waited for her to finish and then framed his answer, looking at her directly.

'My apologies, young lady – I think the applause must have awakened you! I think you'll find, if you ask your neighbours, that I have been saying precisely that for the past hour.'

For a second, Lucy could not take it in. Then, as the horror of what Wilgoss had just said began to register, Lucy could feel her cheeks catch fire and, appallingly, her eyes began to fill with tears. She forced herself not to blink, not to look away, but simply to remain motionless, a hunted animal trying to avoid detection.

The Vice-Chancellor filled the silence. 'Professor Wilgoss, I have a comment myself. Speaking as a non-physicist, I'm intrigued by how much money this basic science must have cost for NASA to launch a satellite all the way to Jupiter to test it.'

As Wilgoss responded, Lucy played over in her mind what she had said. Was it really true that Wilgoss's lecture had included the case of an object moving in a straight line? Well, no, she was sure she hadn't missed that. But should it have been obvious that you could make the jump from accelerating in a circle to accelerating in a straight line? Did Wilgoss mean that it was so patently true that it could be taken for granted? Lucy began to think that this was exactly what Wilgoss had meant, but why shame her like that in front of the whole auditorium?

As the Vice-Chancellor wound up the proceedings, and invited a final applause, Lucy found herself clapping with the rest of the audience, seeking anonymity in conformity.

She stood up with those around her and then had to wait with the others in her row while those in the upper tiers disgorged themselves into the aisles, making their way back up to the exits around the circumference of the amphitheatre.

As Lucy stood there, despondency settled down on her like a dark-grey cloud. Everyone in the audience must have understood the implications of Wilgoss's lecture – otherwise, why did the people standing beside her not offer her a word of solace, of empathy? The familiar feeling of being out of her depth at university descended upon her once more.

Her row began to move into the aisle, and, as she climbed up the stairs to the back of the amphitheatre, she began to wonder if she didn't belong in physics after all. But if she gave up physics now, not only would she be at a loss to decide which subject to switch to, it would mean the end of her childhood dreams of becoming a physicist.

She passed out through the exit door and prepared to walk forward into the anonymous crowd. As she did so, a hand cupped her lightly at the elbow. 'Lucy!'

She spun round, a taught hairspring uncoiling. 'David!' Of all people, she instantly felt ashamed in front of David, he who had been so impressed with her performance in the tutorial and afterwards in the coffee bar. For him to be waiting at her particular exit door, he must have seen her in the amphitheatre – if not before, then certainly when she had broadcast her stupid question to the whole auditorium. She wanted to apologise, to excuse herself to him, but this would only make her feel even more wretched.

'I saw you in the audience,' she stated lamely.

'And I saw you!' responded David, and a wide grin erupted on his face. In Lucy's mind a momentary flash of anger at his mockery was rapidly followed by recognition that this was not ridicule: David was gently teasing her, but he was offering her support.

'Oh, David, it was awful! I'm so embarrassed, I just want to disappear!'

'Hey, wait a minute – I thought this might happen! You don't realise, do you? Your question got to the very heart of the matter, and he never acknowledged it!'

Hope began to break through the cloud.

'Are you saying that my question was OK? That it wasn't stupid – it wasn't obvious?'

'Stupid? Goodness me, no, absolutely not! Wilgoss only gave half the story, and not the fundamental half, at that.'

'Then why did he say what he did? It seemed that everybody in the audience understood him except me!'

'Look, not here, it's too crowded,' said David, looking around. Most of the audience had headed down the aisles to the front, leaving by the doorway on the ground floor behind the lecture platform, but those towards the back, including Lucy and also some of the academic staff, had chosen to leave by the upper rear doors and were emerging from the exits now, and David clearly wished to say something that was not for their ears. At that moment, Professor John Oakhill, Head of Department and Head of Theoretical Physics, walked out of an exit directly towards them.

'Ah, the intrepid young physicist!' he said in an Aberdeenshire accent, looking at Lucy, a broad smile deepening the laughter lines in his face. 'You asked the question that everyone else should have asked, including Wilgoss himself!'

'Oh, Lucy, this is Professor Oakhill,' said David. 'John, Lucy Darling.'

'I take it you *are* a physicist?' Oakhill asked, grinning. 'Only a physicist would have seen through to the root of the issue. But, forgive me, you can only be in the first or second year?'

'Yes, I've just started, Professor Oakhill,' said Lucy, now aware that her cheeks were reigniting. 'I'm hoping to graduate in physics.'

'It's John – please call me John.' Turning to David, he said, 'I hope you'll be taking Lucy to my party? You will be able to come, won't you?' he asked, looking back at Lucy. Before she could respond, a gowned lecturer had taken him off to the side and was earnestly bending his ear.

'That was John Oakhill!' said David, smiling. 'Come on, this is our cue to get away.' As they joined in the crowd flowing towards the stairs, Lucy asked: 'So is he a physics professor?'

'Yes, he's a character, isn't he? He's technically Wilgoss's boss, amongst other things.'

'Was he serious about the party? When is it?'

'Oh, he was serious all right. Every year he has a party in his house for all the theoretical physics lecturers and all the postgraduate students. But he invites others as well. Although it's probably unprecedented that he's inviting a first-year undergraduate!'

'Maybe it's because he thought I was with you. I shouldn't really go,' she went on quickly.

'Oh, Lucy, you must! It would be a great opportunity to meet some of the lecturers informally. And they've always been good parties.'

Lucy was pleased that he was saying she should go, but she wished he had said it was because he, himself, wanted her to be there.

'When is it?' she asked.

'Let's see, yes, it's usually the first Saturday in November, but this time it's a week later. It must be four weeks from this Saturday. You really should come,' he said again.

Maybe he wanted her there for himself, after all.

'Yes,' she said, making up her mind, 'I'd like to go to the party.'

They agreed that David would pass on the details when he got them from Oakhill and then Lucy changed the subject: 'What were you going to say about Wilgoss before it got too crowded?'

'How do you fancy going to the West Gate Bar and I'll tell you there?' David suggested.

She hesitated for only a moment: she dined every night in New Hall where the meals were included in the fees, so that eating out would effectively mean paying twice. 'Good idea,' she said. She thought about adding that she would pay her way but decided that might look too presumptive, as though she might have expected David to offer.

They were being buoyed along by the flow down the stairs and it was all Lucy could do to keep her footing while she risked a sly glance at David's tall profile beside her. Why had he come to her exit? Part of the reason seemed to be that he wanted to put right an injustice, as he saw it. But then, nobody else in the audience had felt moved to come up to her. Well, Professor Oakhill had. Maybe David had felt obliged to come and support her because he felt in some way responsible for her, having saved her life. She hoped this wasn't the explanation.

They spilled out of the entrance door of the Physics Department into a steadily thickening crowd gathering on the wide concrete forecourt outside the building, under the orange-pink glow of the high-pressure sodium lamps. The pavement glistened with evidence of a shower that must have fallen during the lecture. The scene reminded Lucy of crowds emerging from a theatre performance, but she could not escape the conviction that the rising noise level from those around her was more to do with her own disastrous performance rather than that of the newly inaugurated professor.

As they climbed the steps to the five-minute walk into town, she couldn't wait until they got to the bar and so she raised the topic again: 'You were about to say something back there about Professor Wilgoss?'

'Oh, yes, I'm so sorry about that, Lucy,' he said.

'It was hardly your fault! That was nothing to do with you.'

'Guilt by association. He's my supervisor. I must say, he's always come across as condescending, but he's never been so publicly rude before. I was shocked. Actually, I think even the Vice-Chancellor was shocked – did you see the way he tried to smooth it over?'

Lucy smiled ruefully. 'You know, I don't think I remember anything after Professor Wilgoss demolished me.' She was tempted to ask what the people in the audience would be thinking of her – would they enjoy telling their friends about the spectacle? Or would they feel sorry for her – would they perhaps recognise the merit in her question? But she was dissuaded by what remained of her self-respect, and by not knowing if she would really want to hear what David might tell her.

David looked directly at her. 'Lucy, if anyone was demolished this afternoon it was Wilgoss himself! Very few in the audience would have made the jump that you did, to realise that his explanation of centrifugal force meant that the whole universe must be responsible for the inertia of matter in it. But once you said it, the physicists present could see that what you said was true. And so did Wilgoss, that's my point. I've no doubt he knew it already, but when you said what you did, he realised he'd blundered by not exploring that idea in his lecture – and let's face it – that's a much bigger idea than just centrifugal force.'

Lucy turned this over for a moment, and brightened. 'I see. So that's what Professor Oakhill meant when he said Wilgoss should have asked himself the same question!' Then another thought intruded, threatening to unpick the fragile repair that David had begun to apply to her self-confidence. 'So does that mean I've made an enemy of

Wilgoss? I mean, have I shown him up in front of his audience? Even exposed him to ridicule?'

'If he's been exposed to ridicule, then he has only himself to blame. And difficult as it is to understand the logic in Wilgoss's mind, there's no reason for him to pick on you.' He stopped for a moment as they crossed the busy road outside the tennis courts.

'Lucy, I know there's no point in my telling you to forget what happened, because I know you won't. So, instead, when you think back on it, I hope you remember what you said in that packed lecture theatre with pride – I'll bet others in the audience wished they had asked your question.'

Lucy couldn't help laughing. 'Not when they saw what a response it got! But thank you for saying these things – it's just what I needed to cheer me up. I was at rock bottom.'

And she really did feel as though David had lifted her from the depths of her despondency.

CHAPTER 6

Monday 11 October

They were at the West Gate Bar now, and David stood aside to let Lucy go in first. He thought about asking her if he could take her coat but decided that it might look a bit proprietorial.

In addition to the local patrons, the bar was well used by students in term time and equally by holiday makers when the students had gone home. They sat down in one of several small alcoves, each, curiously, supplied with its own flat-screen TV set in the wall, thankfully muted. They both chose the same dish from the menu and David went to the bar to place the order. He returned to Lucy with two glasses of red wine.

'Thank you,' said Lucy. 'What do I owe you?'

'Oh, nothing, it'll be included in the bill – the meal's on me.'

'Thank you, David, but no, you paid last time. I need to pay my way. Sorry, but I insist on going halves.'

'OK, very well then, that's decent of you, thank you.'

'So how was *your* day?' she asked.

'Pretty ordinary until the inaugural lecture,' said David, teasing, and then corrected himself. 'Actually, no, that's not quite true. There was quite an interesting discussion about omniscient gods, and that reminded me that I said I'd tell you about Gödel's Incompleteness Theorem.'

'Is that what you've been working on in parallel with your PhD? Did you say something about *God's theorem*?'

'Hah! Pretty close, actually; it's Gödel's Incompleteness Theorem, or often it's just called Gödel's theorem. Gödel was a mathematician. What he proved was that, in any sufficiently complex mathematical system, there will be true statements that cannot be proved using the rules of the system.' He could feel himself getting excited just discussing the concept. He had not shared his thoughts with anyone, and he wondered now if he were being a little reckless in telling Lucy about his idea. However, it felt liberating to discuss it with someone, and he believed Lucy would keep it to herself if he asked her. He was at once struck by the parallel between this sense of liberation and Lucy's own relief at discussing relativity with someone else for the first time.

Lucy's puzzled face looked cute to David, who burst out laughing. Lucy couldn't help joining in.

'OK, I didn't sell it very well,' he admitted. 'But you should have heard the guy who explained it to me – he was impressive.'

'I'm sorry, I really didn't mean to look blank – but you have to admit, as a trailer, it could stand some polishing.'

David went on to recount his chance meeting with Peter Brown at the party six months earlier, how Peter had told him about Gödel's Incompleteness Theorem and how he had later given him an English translation of the original paper along with commentaries. He told her how he had worked at the paper sometimes for days and nights continuously, sometimes sleeping only a couple of hours, how the theorem had nearly taken control of his life until about a month ago when he had the epiphany that Gödel-seekers talk about – the moment when he finally understood the theorem.

He stopped speaking, conscious that he had been talking non-stop. 'Sorry,' he said. 'I get carried away sometimes.'

Lucy smiled. 'David, I think I've understood most of what you're saying, or at least the essentials. But I'm missing something – I'm not sure I get the significance of all this. What was it that grabbed your attention right from the beginning?'

David paused while the waitress brought the order. She was just Lucy's age: probably a student working nights to defray her debt.

'Ah, I've saved the best for last!' he continued. 'That night at the party, I had what you could say was a Eureka moment. It struck me that if you reduced the universe right down to its mathematics, then, because of Gödel's theorem, there would be things about the universe you couldn't prove.'

'Oh, I think I get it now,' said Lucy, cutting into her steak ciabatta. 'So, for instance, you can prove why the magnetic field round a moving electron is the size it is using relativity, like you showed me, but you might not be able to prove why the electron has the mass that it has, because maybe that's a – what do you call it – a Gödel truth?'

'Yes, you've got it exactly! I like your phrase, *Gödel truth*. The mathematicians call it an *undecidable proposition* if you can prove neither the statement nor its negation, the opposite of the statement.' David took a mouthful of steak and chewed for a minute, waving his knife at Lucy until he could speak again. 'When I went home after the party I looked on the internet and for a while I was bitterly disappointed, because I discovered other people had thought of the same idea. There's a whole Gödel industry out there talking about the fact that the universe is a mathematical system and that Gödel's theorem proves you couldn't have a Theory of Everything that could be completely derived using the mathematics of the universe.'

'You mean that you couldn't prove that the electron mass has to be what it is, for example?' asked Lucy.

'That's right. Although the string theorists hope they'll eventually be able to deduce the mass of particles like the electron from the vibrational patterns allowed by the geometry of the dimensions that the strings occupy. But that's a distraction – if they're right, it would just raise other questions like *why that particular geometry?*' He paused for a drink of wine.

'Anyway, I began to notice that most people who claim Gödel's theorem applies to the universe just *assume* that the universe is described by the equations of the Standard Model – few had actually tackled it formally. So this is what kept me going all those months to understand Gödel's Incompleteness Theorem – the idea of seeing if I could reconstruct Gödel's theorem using the mathematics underpinning the equations of the Standard Model.'

'The Standard Model?' queried Lucy.

'Oh, it's a collection of the equations intended to describe just about everything in the universe – all of the fundamental particles and how they interact with three of the four basic forces. Gravity is the only one missing at the moment, but the string theorists are working on that, too. But, in order to apply Gödel's theorem to the universe, first I had to demonstrate that the universe is purely mathematical.'

'But the universe must be mathematical, ultimately, mustn't it? I mean, it follows a set of rules, although we might not know all of them yet. And that's what a mathematical system is really, isn't it – just a set of rules?'

'Yes, that's right, more or less. The rules could be called axioms – truths that are so basic that you can't prove them, at least, not in their mathematical system.'

'So I'm not sure what more you need to do, if you already accept that the universe is a mathematical system,' said Lucy.

She had a point. David tried explaining it in a different way. 'Well, the way I want to tackle it is to get a feel for how you might write down a Gödel sentence for the universe.'

'You just lost me.'

'OK, remember I said that the crux of Gödel's theorem is that there is a sentence, expressed mathematically, that Gödel used in order to show that a mathematical system is incomplete?'

'Yes, how did it go? Something like: *This sentence cannot be proved in this mathematical system*?'

David was impressed. 'Well remembered! And by *cannot be proved*, incidentally, they mean that, if you start from the axioms, the given truths, then you can't logically deduce the mathematical expression of that sentence from the rules of the system. That's the Gödel sentence.'

'OK, and because you can't deduce it, that makes the sentence true, which means that it *is* a truth that cannot be proved starting from the axioms!' David nodded encouragement. She went on: 'So why can't you just use the same Gödel sentence for the whole universe? Why do you need to get a feel for how to write it down?'

'Well, in the first place, the Gödel sentence wouldn't be expressed in English – it has to be written in the syntax of the system.'

'You mean like writing the symbols two plus two equals four?'

'Yes, that's right. Most of my difficulty in understanding Gödel's theorem was in the convoluted way that he has to use the syntax just to make his famous sentence refer to itself.'

'So what you're saying is that you want to work out the mathematical syntax that underpins the laws of the universe and then see how you could create a Gödel sentence using that syntax? And then, when you finally succeed, overhead, without any fuss, the stars will go out!'

David smiled uncertainly.

'You don't know *The Nine Billion Names of God*? Oh, it's a lovely little story by Arthur C Clarke. These monks in a Tibetan monastery believe that their purpose here on Earth is to write down all nine billion names of God, who will then close down the universe. They hire a computer to write the names faster, and when they have written the last name, the story ends with the brilliant line: *overhead, without any fuss, the stars were going out.* I'll bring a copy from home next time I'm there so you can read it.' Abruptly, she put her hand over her mouth and looked guiltily at David. 'Oh, I've gone and spoilt it now, haven't I?'

David laughed. 'Don't worry, I'd still like to read it. But you know, there's an even more Machiavellian aspect to this. It's a kind of grand cosmic joke! If you now consider that the whole universe is the mathematical system that we're talking about, then you can never be sure that you have created a Gödel sentence! You might write one down that you *think* might be a Gödel sentence, but you can't *prove* it's a Gödel sentence.'

'Hang on, let me just think about that,' said Lucy, and she sipped reflectively from her wine glass. 'Yes, I think I see,' she said carefully, putting the glass down. 'To prove it was a Gödel sentence, you would have to prove your sentence was – what did you call it? – an undecidable proposition. OK, so that must mean you have first to prove the sentence cannot be proved, and, secondly, you have to prove that the negation of the sentence cannot be proved. That's what you said was the definition of an undecidable proposition, right?'

David nodded, smiling broadly, waiting for her to carry on, wondering if she would get there.

'OK, let me think about the negation bit later, if I have to, and concentrate on the first bit, the bit that says I have to prove that the sentence cannot be proved. Oh, but *the sentence cannot be proved* happens to be exactly what the sentence says – it's the sentence itself! So if I can prove that the sentence cannot be proved, then I would have proved the sentence itself, and that would mean I would have just proved a lie! But we're talking about using the mathematics of the whole universe as our mathematical

78

system here. This means that if you could prove a sentence is a Gödel sentence for the universe, you would have used the laws of the universe to prove something that is false, which the mathematics – the universe – can't allow – so that means you can never be certain that a given sentence is a Gödel sentence for the universe!' Her last words galloped out triumphantly, *ta-dah!*

'Goodness, Lucy, you're frighteningly quick to grasp this! I'll bet you wouldn't have taken as long as I did to understand Gödel's theorem!'

She blushed at once and looked down, but she was evidently elated by his praise.

'So, have you made any progress with applying Gödel's theorem to the universe?' she asked him.

'Not as much as I'd hoped,' David conceded. 'The problem is really that there's no guarantee that the Standard Model is the last word even as far as it goes – that's even before you start adding gravity.'

He was now baring his soul to her. He wouldn't have dreamed of opening up the details of his thinking, so recent and undeveloped, to any of his peers. He took a swig of wine and then resumed. 'I began to accept that if you just collect together a number of equations, then you can never guarantee that there are no fundamental characteristics of the universe that you haven't described – in fact, which may not even have been discovered yet.'

Looking at her across the table, he wished he could ask her about her boyfriend. He corrected himself – what he really wanted was to hear was that she had no boyfriend and that the student who had been whispering to her had just been trying his luck without success. But it hadn't looked like that.

'So, recently,' he carried on, 'I've been thinking about changing my approach. Rather than actually writing down a Gödel sentence for the universe, I've been trying to see what the general form of a Gödel sentence for the universe

79

might look like. If you can still prove that sentence is a Gödel sentence, then you must be using a part of the universe that you haven't included in the mathematical description that was used to construct the sentence.'

Lines of concentration troubled Lucy's brow. David thought they looked pretty. 'Yes, I think I get it,' she said. 'We know Gödel's sentence can't be proved within the mathematical system that it is constructed with, so if use what you think is the mathematical system of the universe and you find you *can* prove that some sentence is a Gödel sentence, then that would mean you must have used a corner of the universe, so to speak, that must still be lying *outside* the mathematical system that you had hoped would describe the whole universe.'

'Exactly – and this might tell you where the mathematics needs to be extended. But I have to tell you, I've found this a fiendishly difficult problem, turning physics into pure axioms and rules. In fact, I found out while I was researching the topic that it's one of Hilbert's notorious twenty-three problems – the sixth, I think it was – and still unsolved. Oh, David Hilbert was a famous mathematician who published the problems at the turn of the new century, nineteen hundred.'

Lucy considered this. 'How would you account for the human mind in your mathematical description? What I'm thinking is – surely the brain will always be able to apply Gödel's theorem, even to the universe. So that kind of implies that it can somehow stand outside any mathematical model of the universe, don't you think?'

'I know what you mean,' said David, 'but then you would have to believe that the mind, or consciousness at least, is literally something supernatural for it to stand apart from the universe.'

'So how would you see the mind being described mathematically, then?'

'Well, of course, I couldn't give you chapter and verse, but it would need to go something like this, I think. You have to realise that the mathematics of the universe will operate fairly simply at a fundamental level, but that a fantastic range of richness and complexity emerges from these simple rules. Anyway, the properties of cells and living systems ultimately derive from these simple rules and they would include neurons in a network that can respond to sensory inputs and can make models using these inputs. Sophisticated networks can even model themselves making such models – they are self-aware. Of course, it's not obvious starting from the basic rules that all of that follows!'

They both laughed.

'That's interesting,' said Lucy. 'I'd never thought about self-awareness so – mechanistically. I'll have to think about that in the quietness of my room tonight.'

David pictured her in the quietness of her room.

'Funny enough,' he said, 'we touched on that in coffee this morning. Oh, and something else I was saying just now reminds me of the same conversation this morning. It's to do with not being able to identify a Gödel sentence for certain, when we're talking about the complete universe.'

'Mmm,' said Lucy, nodding, her mouth full.

'Well, we were arguing about whether the universe could have an omniscient being. Such a being might know an awful lot, but there would be some truths in the universe – and I'm talking Gödel truths here, of course – undecidable propositions – that it would never know for certain were indeed truths.'

He skewered the last morsel of his steak and began to chew it, waving his knife again until he could continue. 'You said last time that you would expect me, as an atheist and a scientist, to have a pretty convincing proof that there is no god.'

'Ah, I think I see where this is going!'

David smiled. 'You're probably right. This supposedly omniscient being could not know, for instance, whether Goldbach's conjecture was true or not. Assuming, of course, that Goldbach's conjecture is an example of an undecidable proposition, but you could pick any undecidable proposition. I'd expect the Judeo-Christian God to know the answers to such things. So Gödel's theorem seems to me to rule out at least an omniscient god.'

'I'm not sure I remember what Goldbach's conjecture is, but I get the idea – there will be truths that an omniscient being wouldn't know whether they were indeed truths or not.'

'Oh, Goldbach's conjecture is that every even number is the sum of two prime numbers. Like twelve is the sum of five and seven, for example. Nobody knows yet whether the conjecture is true or not, because nobody has ever been able to prove it. My guess is that it will be proved true eventually, but it might genuinely be an undecidable proposition. Actually, it's a good example of the knots you can get into when you play with the kind of logic that underpins Gödel's theorem. Because you just need one example of an even number, no matter how big, that is *not* the sum of two primes and you would have disproved Goldbach's conjecture. So if it is possible in principle to *prove* that Goldbach's conjecture is undecidable, then actually you can conclude that Goldbach's conjecture must be true!'

Lucy was puzzled. 'How come?' she asked.

'Because if Goldbach's conjecture is false, then it is possible to *prove* that it is false, in principle, just by going through all of the even numbers and testing each one to see whether it's the one that can't be made up by summing two primes. It might take a long time, but eventually you'd come to it.'

'Oh, I get it now – if Goldbach's conjecture is false, then it wouldn't be undecidable because you could definitely prove it was false, given enough time. So, if it's really undecidable, it's not false, so it must be true!'

'Yes, so, to sum up, if you can prove Goldbach's conjecture is undecidable, then you have proved that it is not false, or, in other words, you have proved it is true. But if you can prove something is true, then it isn't undecidable! So you can't prove that Goldbach's conjecture is undecidable! Although it may be!' he added, to Lucy's amusement. 'Anyway,' he went on, 'you're right, it doesn't matter which Gödel truth you pick on – we just know there will be some in the universe – actually, an indefinitely large number – and yet the god won't know if they are really true or not.'

'I'm sure there has to be a flaw in your reasoning about there not being an omniscient god,' said Lucy, 'but just because I can't spot it right now doesn't mean I give up.' She smiled. 'I'll have to add that to my list for my room tonight.'

The waitress cleared away their plates and asked whether they wanted pudding. They both agreed just to have coffee.

Lucy changed the subject. 'Can you tell me about your PhD? Will your research be helped any by all the time you've spent on Gödel's theorem?'

'Sadly, no. At least I don't think so. My research topic is to do with the holographic principle. Have you heard of it?'

'I know what a hologram is, but *holographic principle* sounds rather more fundamental.'

'Yes, it is. The whole thing is very debatable, and I'm not sure I believe it, but the idea is that all the information inside a black hole, or the universe, for that matter, is actually encoded on the black hole's surface.' David knew he wasn't projecting enthusiasm for his research, but he couldn't help himself. 'Anyway,' he went on, 'this raises all sorts of questions, like – is the information on the surface like a true hologram, so that any area contains some information about the whole universe, and what happens if the surface is expanding, like in our own universe? That's my topic. The last question, I mean. I'm looking at how the surface encoding would affect the information inside the universe as a consequence of the surface expansion.'

The coffee arrived and they both added milk. Then they each took two large lumps of the Demerara sugar and began to stir it into their cups. Simultaneously they both said 'Snap!' and then broke into a giggle at the unexpected synchrony.

'You didn't sound very excited by your PhD topic,' Lucy observed. 'Much less than you were about your ideas on Gödel's theorem. I hope you don't mind my saying that?' she added.

'No, you're quite right. I was given the topic by Jeremy Wilgoss – it's not one I'd have chosen myself.' He was suddenly conscious that he had been doing most of the talking. 'Look,' he said, 'I've spoken a lot about what I've been doing, and what I want to do and I'm probably at risk of boring you,' and he half raised a conciliatory hand as she was about to protest, 'but I'm curious to know what *you* want to do in the long-term. I mean, are you thinking of research?'

'My first goal is just to get my degree,' she said, 'and of course I'd like to get the best degree I'm capable of. So I'm not going to waste my time while I'm here. I don't want to leave university thinking that if only I'd worked a little harder I'd have done better.'

'Let's say you get a First – then what?'

'If I were that lucky, then, yes, I'd love to do research. I guess I'd start with a PhD, like you. But that's so far into the future!'

'Not from my perspective,' said David, and mentally kicked himself. He must avoid drawing attention to the difference in their ages, a difference which, to him, might seem slight, but which would appear significant to Lucy, an asymmetry in perception into which he was only lately becoming inducted, as the opportunities to take the older view increasingly presented themselves with the passing years. He covered up: 'What aspect of physics would you like to do research in? I'm guessing you'd like to do it in theoretical physics?'

'Yes, absolutely. In fact, come to think of it, you're my role model!'

To his consternation, David could feel himself blush, flattered that she should think so highly of him, disconcerted by the fact that a role model is commonly much older than the disciple.

'Did you come here as an undergraduate?' Lucy asked him.

'Yes, I thought it was the perfect combination of an ancient university and a town which is a tourist attraction in its own right.'

Lucy thought so, too, and then they started to discuss the staff who taught the various first-year physics courses, with David giving her the low-down on some of the lecturers. When they had finished their coffee, David said: 'It's Monday night – I guess you'll be anxious to get back to review the day's lectures?'

Lucy agreed and so David paid the bill and she gave him half. David left the waitress a generous tip.

As they stepped out onto the pavement, slick and shiny with another shower of rain, the street lamps multiply reflected in the protected token tracts of cobblestones smoothed under the tread of generations, the atmospheric interplay of street and night and rain and light blending into an impressionist painting, David was, all of a sudden, downcast by the looming prospect of this wonderful evening drawing to a close.

He offered to walk Lucy back to New Hall.

'Oh no, you don't have to do that,' Lucy protested.

'Please, I would like to,' he insisted, gently.

They retraced their path, then carried on beyond the Physics Building towards New Hall, talking all the while of similarities and differences between the current course structure and what it was like when David had taken it. Throughout their walk back to Lucy's hall of residence, David kept wishing that he could think of a way to ask her about the question that preoccupied him – was that student in the tutorial room her boyfriend? With each step they took, he could feel the opportunities slipping away like sand through an hour-glass until, finally, they arrived at the entrance to the hall, the end of the line.

Lucy turned to look up at him. 'Thank you, David, for rescuing me tonight. That's twice – it's getting to be a habit!'

David smiled and started to reply, but she continued: 'Shall we meet after the tutorial on Friday?'

For a second, David was not sure he had understood her, and then his heart lifted. She was making a date! Never in his life had a girl made a date with him! The next instant, his momentary joy was extinguished by drenching disappointment.

'Oh, damn,' he blurted out, showing more of his despair than he meant to. 'Some trustees of the Wane Ringer Cade scholarships are visiting on Friday afternoon and I have to meet them – it's a three-line whip, because I'm a scholarship recipient.' He looked at Lucy, hardly daring to hope, and heard himself say: 'What about Saturday – do you think we could meet then?'

David thought he saw a flicker of hesitation in her eyes but then she smiled. 'Yes, that would be good,' she said. 'We could meet in the Maidenhead Café, say after lunch, at two o'clock?'

'At the far end of North Street? Yes, I'll look forward to that very much,' he replied, giddy with the speed of events, desperately trying to think of something else to say to prolong the moment, failed, and finished lamely, 'Well, see

86

you on Saturday.' Two students emerged from the foyer, triggering Lucy to take a step back towards the door, and David waved. 'Bye, Lucy'.

'Bye, David,' and she turned into the entrance as he set off back towards the town.

CHAPTER 7

Friday 15 October – Saturday 16 October

'Got any fundamental questions to ask today?'

Lucy's soaring spirits crumpled in mid-flight as she recognised George's voice too close within her personal space behind her. They were walking along the corridor towards the tutorial room where she had been looking forward to seeing David again – not necessarily to speak to him, but just to see him, to listen to him, to watch the way he moved, confidently but not arrogantly, not like George, she thought. George had been trying all week to make her take notice of him, but she had generally managed not to look him in the eye, keeping her attention carefully focused elsewhere when he was in the vicinity. She reproached herself for letting her guard down this time, coming into the corridor without first checking on his whereabouts.

She halted and turned to him abruptly, hoping that her aggressive reaction would offset any concession he might read into the fact of her stopping. 'What do you mean?' she asked, trying to neutralise her tone, to remove any possible traces of friendliness in the question.

'Well, two fundamental questions in two weeks – it must be time for the next one, surely?'

So he had heard about the inaugural lecture – damn!

'I ask the questions when they occur to me.' She decided not to elaborate, to get involved in a conversation, and she made to turn and resume her journey.

'Yes, I'm sure, but you must admit, it gets you noticed. Where do you get them from anyway, these questions?'

Lucy was momentarily speechless, scandalized that he should think she was cribbing questions to get attention. She forced herself to reply calmly, not to give in to the release of anger into which her pumping adrenalin was threatening to push her.

'Why do you think that I must be getting them from anywhere?' she said, her voice still studiedly neutral.

'Well, how can I put this delicately? We think differently, men and women. Women are better at languages, at personal relationships, they have higher emotional intelligence. Men think spatially and laterally – that's why they're good at physics. Men make the wild leaps of imagination that take their subject forward. They ask the kind of questions that you were asking, that's all.'

Lucy was incensed. She was so annoyed that she decided not to dignify his arrogance with reasoned debate. 'So when you encounter a woman thinking laterally and making wild leaps of the imagination, you just don't believe it? Well, as they say – *get over it!*' With that, she turned smartly on her heel and walked off before he could retaliate.

She sat in her usual seat at the back of the tutorial room, still fuming, gratified that George, who had chosen a seat in the front, appeared at last to have understood that his attentions were unwelcome, although she could not be completely sure of being undisturbed until David appeared, pre-empting any last-minute change of seating arrangement on George's part.

According to form, Lucy was an attentive but silent participant throughout the tutorial, until the very end. David had decided to set a problem for the class. He explained that there was no compulsion to tackle it, but that he felt that it might be a useful exercise for those who wished to attempt it.

'Suppose we dig a tunnel right through the Earth,' he said, 'starting at, say, the North Pole and finishing at the South Pole. First, though, you've got to assume that the Earth is solid all the way through. Oh, and you'd better assume that its density is the same throughout, as well.

Now, suppose I travel to the North Pole and drop an apple into this hole. It will fall all the way to the centre, of course, where it will be travelling so fast that eventually it will have gone all the way through the tunnel to the South Pole. Oh, and you've got to assume no air to slow it down. The problem is to find how long it will take the apple to fall all the way from the North Pole to the South Pole.'

There was a ripple of amusement as the students pictured this.

'Out of interest,' David continued, 'would anyone like to make a first guess at how long it will take the apple to travel all the way?'

Lucy was fascinated with the idea. She focused in on the apple, falling through the hole, the gravitational force on it diminishing as it approached the centre of the Earth. Immediately, she realised that she would only have to work out the force from the sphere of matter that remained below the apple at any time – the force from all the matter in the spherical shell above it would cancel out completely.

As though from a long way off, the rumble of conversation among the students that only just penetrated her concentration was interrupted by David pointing to a raised hand. It was George's.

'Without giving it too much thought,' he said, 'it must be in the order of days rather than weeks.'

'OK,' said David. 'Any other suggestions?'

Without warning, the tumblers lined up: she had the key to it. She raised her hand.

David pointed to her, not calling out her name, she noticed, which was good. She didn't want the other students to think she was being given any special favours by her rescuer.

'I think it would take just a little less than three-quarters of an hour,' she called out, to some tentative laughter.

'Amazing! Yes, it would take forty-two minutes,' said David. 'How did you do that?'

'Well,' she began, 'when the apple is at a given distance from the Earth's centre, you only have to think of the force from the sphere of rock beneath the apple, because all the forces from the spherical shell of rock above it cancel. So the force is proportional to the mass of the sphere below the apple. That means it's proportional to the cube of the apple's distance from the Earth's centre. But from the law of gravitation, the force is also proportional to the inverse-square of the distance from the centre, so the overall effect is that the force on the apple is proportional to its distance from the centre of the Earth.'

Before she had raised her hand, she had not worked this out in detail: it was more that she had visualised the sphere with its radius cubed and the inverse-square behaviour of the gravitational force and she was only now working it out as she described it to the class. To her relief, though, it came out to the same answer that she had seen in a flash.

'Well done,' said David. 'But how did you get the time of travel?'

'Well, with the inward force proportional to the distance from the centre, it meant that the apple just oscillates like a simple pendulum back and forth between the two poles. But then I realised that if you think of a satellite – an apple – in low-Earth polar orbit, then the force of the Earth on the apple-satellite – in the North-South direction only – is proportional to the distance above or below the equatorial plane, just like the force on the apple in the hole.'

'Is it?' said David. He sounded surprised. 'Why?'

Again, she had not worked it out in detail, but she had pictured it almost immediately after she had seen the connection between the force on the apple and its distance from the centre of the Earth. She hoped that it would work out again as she had visualised it.

'Because the total force on the apple-satellite – its weight – always points to the centre of the Earth,' she began, 'and so the component in the North-South direction

91

is equal to its weight times the sine of the angle that the radius to the apple-satellite makes with the equatorial plane. But its distance from the equatorial plane is also the radius times the sine of the angle. So the force on the apple-satellite in the North-South direction is proportional to its distance from the equatorial plane.' Good! It was coming out as she had hoped.

'In other words,' she went on, 'the component of the apple-satellite's motion in the North-South direction must be identical to the oscillation of the apple down the hole through the centre of the Earth, since they both have the same force towards the centre of the Earth when they are at the North Pole. And I knew that a satellite in low-Earth orbit takes ninety minutes to make a complete circle, so it would take forty-five minutes to go from the North Pole to the South Pole. But low-Earth orbit is a few hundred kilometres above the surface of the Earth, so the time would be a little less than three-quarters of an hour.'

'Well, I've never in my life come across an account of the motion of a satellite in terms of its distance from the equatorial plane,' said David, 'but, of course, you're perfectly right, now that I think about it. The orbit can be entirely described in terms of simple harmonic motion in that direction and in the direction at right angles. Still,' he continued, 'I think it would be useful for the class to work out the solution for themselves. They don't have to do it your way – that has to be the biggest leap of lateral thinking I've come across first-hand,' he added.

Through her embarrassment, Lucy glanced at the front row, and found the back of George's head. As a rule, retribution was an emotion alien to her soul, but this was a delicious exception. *Gotcha!* she thought.

*

92

The next morning was Saturday, and Lucy lay on her back and stretched languorously, luxuriously, diagonally in the double-sized New-Hall bed, the nooks and crannies of her consciousness gradually lighting up like the steadily brightening sky of an advancing new day. In her post-slumber bliss she tried idly to induce a dissociative state, to recreate the pleasant dreams whose mere shadows now remained, the detail already lost in the receding night, but an uneasy gossamer shade kept brushing its butterfly wings on the corner of her mind, threatening to invade her cocoon, to introduce a dissonant note of reality into the peaceful state of her soul.

There! In trying to ignore the intruder, she had focussed on the very source of her slight disquiet – her rendezvous with David later in the day. This was more than just seeing him after the tutorial to carry on a conversation about Gödel's theorem: this was a date! By arranging it on a Saturday they had removed the prop of the lecture room, the academic pretext for a meeting, and taken it to a more personal level, a level that she had never intended to enter because of her commitment to her studies, a level where different rules applied, and from where it would be more difficult to withdraw.

In the fraction of a second after David had suggested Saturday, these thoughts had flitted across her mind. In agreeing to Saturday she had reacted more from instinct than from a rational consideration of the issues, anxious not to appear prevaricating. Why was she now so concerned? After all, had she not been utterly dismayed when David had seen her apparently flirting with George? Surely her reaction indicated to herself how she truly felt about David?

She swivelled her eyes around the room without raising her head from the bed, taking in the desk with her recharging laptop, the Anglepoise lamp and the coffee maker next to it, the shelf that held her rapidly growing library, the Claude Monet prints that she had mounted on the walls with tacky putty, probably in contravention of hall rules, her wardrobe, neatly shut, chest of drawers, ditto, and the twelve-inch wall-mounted television provided, like the double bed, more with an eye to the out-of-term conference and holiday business than for the exclusive comfort of the students. Her window faced due east, and from the angle of the sunbeam now streaming through a chink in the curtains she judged the time to be nearly nine o'clock. She turned her head to check the alarm clock on the shelf by her bed, blessedly muted for the weekend, and saw she was about right.

OK, it was agreed, she did like David. But what about her resolution not to get involved with boys (or men), at least not until she was sure that she could afford the time? Well, she had completed three full weeks' lectures now, and, if she were honest, she had not found the work taxing. The studying that she had done over the previous two weekends had been peripheral to the course – she was having no difficulty in keeping up each evening with her physics or, indeed, her maths, information technology and astronomy modules.

She could feel her resolve melting. For a man like David, she would have the time. In any case, David himself would be as concerned about the need to spend time on physics as she was – at least until he finished his PhD. Suddenly she interrupted her daydream as she caught herself thinking about herself and David so far into the future. This was absurd! In any case, he was six years older than she was – somehow, her thoughts almost seemed illicit! And surely he would already have a girlfriend? Maybe he shared his flat with his partner!

This raised another question. If he was interested in her, why had he not offered to kiss her when he had walked her back to the hall on Monday night? The occasion was right and it would have been perfectly natural for him to have given her just a little peck on the cheek. Perhaps he was

94

gay? That would be just her luck! But then, why had he asked her out?

Of course, it had actually been she who had asked him out first. She had made the first move in proposing that they meet after the tutorial, a perfectly reasonable suggestion, considering that she had said she would like time to answer his conundrum about the impossibility of an omniscient being. During the rest of the week she had spent several hours on the computer finding out what she could about Gödel's theorem. She had been surprised at the vast amount of information that was available about a topic she had not even been aware of until now. If she was going to counter David's argument, she would need to understand at least the elements of the theorem.

Distantly, she registered movement within the hall: students were beginning to surface after their weekend lie-in, but the noises were not intrusive and she continued to gaze at the ceiling as she summarised what she knew about Gödel's theorem from both David and the internet. In a reasonably sophisticated mathematical system, you can prove a vast number of statements, stated in the language of the system, all of which are true, like two plus two equals four. So far, so good. But Gödel had proved that there are some statements that are true but can't be proved using the mathematical logic of the system. He had invented just one particular statement to demonstrate this – his *sentence* – although there was apparently an indefinitely large number of such true statements.

OK, she thought, if the mathematics of the system can't prove that Gödel's sentence is true, how do we know it really *is* true? Answer – because we can stand outside the system and use our logic to see that the clever sentence *must* be true. That was the key, she realised in a flash. You always had to be able to stand outside the system to prove that a Gödel sentence was true. So, what if the system were the whole universe, including ourselves? Then nobody would be able to prove the Gödel sentence true, would they, because nobody could stand outside the universe to prove it. But suppose there *were* a Gödel sentence in the universe – something that was true but couldn't be proved? Didn't that mean that there *had* to be a higher level – *outside* the universe – where you *could* prove the sentence was true?

Lucy was excited now. She had grasped what the theorem said, if not the details of the proof. But she hadn't come across any website saying that if a system contained – what had David called it? – an undecidable proposition – then that would mean there must be a higher system above it in which it *could* be proved. It would be a kind of *inverse* Gödel theorem, where you started from the Gödel sentence and then proved that there had to be a higher logic, a higher level that proved the truth of it. She hadn't seen such an inverse theorem anywhere, but surely it made sense? Surely, whenever a system contained a true statement that it couldn't prove, then the only way for the statement to be true would be if there were a proof in a higher-level system? *Otherwise what was it that made the statement true every time it was tested?*

She wondered if David had come across the Inverse Gödel Theorem (she was already calling it this in her mind). Maybe the theorem was obvious, but had anyone actually proved it? That might be tricky if it were to apply to the whole universe, because then the proof would have to be constructed within the same mathematical system that contained the Gödel sentence rather than in the logic of the higher level in which Gödel's original theorem would be proved.

Her mind was just starting to get tangled into a strange loop when she was jolted upright by a loud thump on her window. Her first thought was that a student was outside trying to get into her room. She bolted out of bed, ran to the window and flung open the curtains. She stood there with her arms outstretched, still gripping the curtains, heart

96

hammering, not moving, just staring out at the sunny blue sky, the grassy banks and the Chemistry Building across the road. She was on the third floor: it couldn't have been a student unless he was Spiderman! As she dropped her arms she noticed a faint white splatter on the window, as if a window cleaner had tried to remove a seagull's dropping and left a residue.

She looked down at the lawn, thinking that maybe a student had thrown something at her window, but the area was deserted, apart from half-a-dozen of the ubiquitous rabbits that grazed, semi-tame, throughout the campus. She undid the latch and pushed on the window. It was hinged at the top and opened out from the bottom, and she gently inserted her head into the gap and looked directly down. The quadrangle of lawn below was bounded on two sides by the L shape of the building – her east-facing wall and, to her right, the north-facing wall of the hall, and by a road and a path on the other two sides. For a moment she could see nothing untoward and then she spotted it. One of the shapes that she had initially taken for a rabbit was actually a largish bird lying on its side on the grass a metre from the bottom of her wall. She thought she could see it move, but it made no effort to right itself. She stared, hypnotised, replaying what must have occurred, the bird flying towards her window, towards the reflection of the sky in the glass, towards what it thought was free air through the frame of her window, the sudden neck-breaking crash, falling to the ground before the poor thing could even be aware of the collision.

Instantly energised, she strode over to her chair, pulled on a jersey and her weekend jeans over her pyjamas, shoving her feet into slippers – no need for her key as it was a deadbolt lock – running out of her room, along the corridor and downstairs to the fire door, pushing through it and dashing round to the neatly cropped lawn. She came to a halt two metres from the bird. It was a pigeon, as far as she could tell (she had never been good at bird recognition, but with a pigeon, she felt she was on fairly safe ground). It hadn't shifted its position since she had looked out of her window. She saw now that the movement she had detected was no more than a fluttering of feathers in the breeze. Its eye was shrouded over with a semi-translucent membrane; the bird was clearly dead.

As she crouched down beside it, she noticed another pigeon walking on the grass in a head-bobbing arc around them, first to the right, then to the left and then back again. The two had been a devoted pigeon pair, and now the one that had been left behind was waiting for its mate to recover so they could fly away together. She pictured the bird just five minutes earlier soaring into the sky, happy in its own fashion, loved by its mate, watched by its mate as it fell to earth already dead. She felt the tears well up in her eyes, overflowing, streaming down her cheeks; she cried for the bird and for its mate, not knowing, not capable of understanding what had happened to its companion, cried for her own loss, too.

Did God feel the same sorrow when someone died? Unbidden, a fragment of St Matthew's gospel, retained from her Sunday-School days, unspooled itself in her mind. *Not one sparrow falls on the ground apart from God's will.* Why would God will the bird to die? What would be the point? As she looked down at the bird she wondered if it had a soul. Did God preserve its memories, its primitive feelings in the same way that He preserved the human soul? The biblical text reminded her of God's omniscience. In a way, she was like God looking down on His people, all-seeing, all-knowing, understanding what His charges could never grasp. From her vantage point it occurred to her that just as she was in effect at a higher level than the bird, looking down on its universe, so God was outside the human universe, looking down from His higher level.

98

Despite her emotion, it dawned on her that this was the answer to David's conundrum. God couldn't be omniscient if He were of this universe – David was correct there. But, of course, God was outside the universe: why hadn't she thought of that and said so in the bar? In a flash, she saw the connection between this idea and her Inverse Gödel Theorem: if the theorem were true, if there was always a higher level above systems that contained unprovable truths, then, if the universe contained an unprovable truth, there would be a higher level to it, too – a higher universe, a meta-universe from where God could see and know and understand everything in our universe!

She stared at the bird and shivered slightly – not entirely as a result of the chilly October morning. Maybe God did have a point in letting the bird die after all! She decided to drop that line of thought right away. She looked up at the matrix of windows facing the quadrangle. Not a soul returned her gaze: in most of the rooms the curtains were still drawn against the light of the morning sun. With a shuddering sigh she slowly rose to her feet, wiped her cheeks with the back of her hand, walked back around the corner, headed for the main entrance and reported the dead bird to the porter. Then, her mind an odd jumble of sadness for the bird and its mate and excitement over the implications of an Inverse Gödel Theorem, she mounted the flights of stairs to her floor.

*

Lucy checked her watch as she entered the cobbled section of Market Street: only a quarter to two and the café must be less than half a mile away – she was still allowing too long to reach places in this small town. She slowed her pace and turned to look at her reflection in the window of a shop selling Christmas bric-a-brac, the gloomy interior helpfully enhancing the contrast of her image. Deciding what to wear for the afternoon had been difficult. At weekends she had been dressing down with jeans and T-shirt; on the other hand, David had never seen her dressed so casually. However, if she wore a skirt, might that look too dressy for a Saturday, as though she were making an effort for a date (setting aside the fact that, technically, of course, this *was* a date)? She had finally chosen her navy pleated skirt with white polka-dots and her white jersey. On top, she wore her *pièce de résistance*, a dark-grey coat, tapered at the waist with a flared skirt and cream faux-fur trimming, secured for a knock-down price on eBay.

Inspecting her ensemble, she was pleased with the result. She certainly stood out among the passers-by reflected from the deeper recesses of the shop window. Yes, maybe it did look as though she had dressed for the date, but she had to admit the figure-hugging outfit emphasised her femininity, lending her confidence. She resumed her stroll up the street, pausing to look in shop windows in the cobbled square, finding herself unexpectedly cheered by the sight of a young couple sauntering across the square, arms crossed behind their backs, hands clasping each other, catching herself smiling benignly at a set of twins in a pushchair parked on the pavement, their mother scrutinizing the dresses on display in the women's fashion shop, her children ogling the freezers in the ice-cream shop next door.

The upwelling of benevolence to the world surprised her – surely she couldn't be in love? How could she be – she had only spoken to David twice – but she had to concede that there did seem to be a connection between her feeling of well-being and the prospect of seeing him in the next few minutes. Unconsciously, she had increased her pace again and she arrived at the T-junction with Castle Street with ten minutes to spare. The Maidenhead Café was just round the corner, but there were no shop windows now to delay her progress and so she carried on directly to the café.

As she reached the door she noticed a sign hanging on the wall outside advertising that the coffee shop was also a bookstore. On entering, she was struck immediately by the cheerful yellow walls, the friendly muffled hubbub from what appeared to be a largely student clientele, and, sure enough, a separate section displaying books that could presumably be thumbed through before buying while drinking coffee. David was seated at a table directly beside a window, facing the door as she came in, and he stood up to greet her with a big smile. Never before in her life had she been met by a man standing up to welcome her and she couldn't help grinning in return.

'I'm early,' she said, 'but you still beat me! Have you been waiting long?'

'Not really. I came early to book a seat and I bought this and started to read it, so the time has passed quickly.' Lucy glanced at the paperback on the table in front of him, a bookmark protruding from the early pages where he had closed it face up. Even upside-down, she recognised the picture.

She was mildly surprised. 'You're reading *The Cabin*? You know that's a religious book?'

'Yes, that would be because this is a Christian bookshop,' said David, amusement playing at the corners of his mouth.

'Oh God, is it? I'm sorry! It wasn't intentional.' She had suggested the Maidenhead Café, not knowing that it was also a bookshop, let alone one devoted to Christian literature, knowing only that it was one of the many coffee shops in town frequented by students. 'You must think I'm trying to convert you!'

David laughed. 'I did wonder, but you said you weren't Christian, so I reckoned we only came here for the coffee.'

'So what do you think of it so far?' Lucy asked, nodding towards the book.

'It's intriguing – the hero's received a letter, apparently from God, to meet him in a cabin he knows. That's as far as I've got.'

'Actually, I've read it,' she confessed. 'Without giving too much away, the author tries to interpret the Holy Trinity for us. It's controversial among some Christians – they think the author doesn't lay it on heavily enough about us deserving punishment for our sins – all the dogma that alienated me from the Christian theology. What kind of coffee would you like?'

'Oh, let me—'

'No, it's my turn. You can get on with your book while I'm getting it. What would you like?'

'OK, thank you, Lucy. Could I have a cappuccino, please?'

Lucy went over to the counter and returned with two frothing cappuccinos.

'So I take it *The Cabin* didn't change your life then?' David asked her.

'No, if anything, it just reinforced my convictions about dogmatic religion.' She smiled: 'Let me know if it changes yours!'

She sipped at some of the froth on top of the coffee. 'Are you making any headway with a Gödel sentence for the universe?' She paused. 'Goodness, how's that for a conversational gambit?'

'Leaves the rest standing! Anyway, not much progress, I'm afraid – everything I've tried recently just seems to hit the buffers.'

Despite his smile, he sounded a little downcast, and she had a sudden urge to enfold him in her arms, to comfort him. Should she tell him now about her idea? She could feel her heart beating faster in her excitement, the anticipation of surprising him with her Inverse Gödel Theorem, and his gratitude for helping him over his mathematician's block, but then a cautionary voice at the back of her mind spoke up to warn her to tread very carefully. He had been working hard on his idea for six months, and so he would feel an ownership, not just of his project, but probably the whole territory surrounding it. If she, the ingénue, came stomping in with her hobnail boots to rescue him, not only might he reject her idea, he might resent her very attempt to help. Then she saw a way forward.

'David,' she began, 'there's something about Gödel's theorem that I haven't quite got my head around yet.'

'Hah! I'm not surprised! It's like a hallucinogenic drug – it messes with your mind. Gödel was quite mad at the end, you know.' His voice had regained some of its confidence: he was teaching again. 'What exactly is it that's bothering you?'

'Well, you know how the Gödel sentence has to be true even though you can't prove it in a mathematical system?'

'Yes, that's right,' he said encouragingly, taking a teaspoonful of chocolate-capped froth and supping it up.

'Well, suppose you were a being that lived entirely within the mathematical system instead of out here. How would you then go about trying to prove the truth of the Gödel sentence?'

'Oh, now that's a good question. A typical Lucy Darling question!' She gave a shy grin as he continued, thinking aloud. 'Let's see. Well, suppose we take these coffee cups. Yours contains all of the statements – the theorems – that *can* be proved. An example would be two plus two equals four. Mine contains all of the statements that *can't* be proved. An example would be two plus two equals five. Can't be proved in the mathematical system, I mean. Call it system X.'

'Right,' said Lucy. 'So you could say that my cup contains true statements and yours contains false statements.'

'Well, in ordinary conversation, yes, but we need to be careful here, because we'd have to say what we mean by truth, then. Philosophers are still arguing about that after three thousand years! We should try to do this without depending on the concept of truth as much as we can.'

'OK, but why do you say that?' she asked.

'Well, cutting to the chase, we know from the result of Gödel's theorem that his sentence can't be proved, not in system X. So, in the end, we know we're going to find the sentence in my cup, which – yes, you're right – normally contains false statements, but, in the case of Gödel's sentence, which we know to be true, my cup will contain that true statement. *We* can see it's true, of course, from our lofty perspective, but they can't in system X, otherwise they would be able to prove the sentence, too.'

'You're right – this really does mess with your mind!' Lucy exclaimed.

'Stick with it – there's more to come! Sorry, I must have a drink of unprovable statements,' and he paused while he swallowed some of his cappuccino. 'Now, let's see: in this system X, they have a few rules. One of them is that if they can prove a theorem, not only do they know they will they find it in your cup, but they know that its negation must be in my cup. So, when they find they can prove a theorem labelled A which states *two plus two equals four*, they know A will be in your cup, and that the *negation of A*, which means *two plus two does not equal four* , will be in my cup, because that can't be proved, of course.'

'Yes, that makes sense. Any other rules?'

'Well, it almost goes without saying that they can't have the same statement in both cups, because that would be like having a statement equal to its negation.'

Lucy nodded. She had not anticipated that her question would lead to such a closely reasoned analysis, but she still had a hunch that it would lead David to the Inverse Gödel Theorem in the end and, anyway, she was finding that grounding the discussion in examples helped to make the concepts more tangible.

104

'Now for the dénouement!' David was enjoying himself. 'One day, the citizens of system *X* wake up to find a grammatically perfect sentence which is labelled *G* and states *G cannot be proved*. That's the best I can come up with for a simple English expression of how the Gödel sentence is made self-referential in the mathematical system. It's a slightly more formal way of saying *This sentence cannot be proved*. Anyway, they don't even know how to start proving it, and so someone suggests just looking in Lucy's cup to see if *G* is in it.'

Lucy took up the thread. 'Right, let me to do this slowly now,' she said, concentrating. 'Say they find *G* in my cup, that will mean that *G* can be proved. But *G* states that it can't be proved, or, to put it another way, *G* states that it goes in your cup. So, if they find *G* in my cup, they know that it must also be in yours!' She hadn't realised she had been tense until she trumpeted the final words and found herself falling back in her seat, suddenly relaxing, looking up, beaming, pleased with herself, not minding that David could see.

David grinned back. 'Yes, and that's against the rules. They can't have the same statement in both cups. So they know they won't find *G* in your cup and so they try looking for the negation of *G* in your cup instead.' He nodded to her to take over once more.

Lucy leaned forward again. 'Right, if they find the negation of *G* in my cup, that would mean that my cup contains the statement *G can be proved*.' She raised her eyebrows at David just to check that she was on the right track, and he smiled back his encouragement. 'OK ... well, that means that, by the rules, if they find the negation of *G* in *my* cup, they will find the negation of the negation of *G* in yours. Which presumably means that they will find *G* in yours?' She checked again with David, who vigorously nodded his agreement. 'But hang on,' she exclaimed, 'in my cup it says that *G* can be proved, which means that they

105

have proved that G can be proved, which means that they will find G in my cup. But we just found G in yours as well!'

'Hah! Yes, exactly! They can't deal with Gödel's sentence without breaking their own rules, rules which seem entirely logical and reasonable. *We* know, of course, that they will find G in my cup but they won't find its negation in yours. But that breaks their rule about a theorem and its negation being paired off into each of the cups. *We* can see the answer because we can look down on them from above!'

'Like God?' Lucy said mischievously, and they both laughed.

'Has anybody ever tried to sort out this conundrum?' she asked.

'Oh, Bertrand Russell and others were aware there was a problem long before Gödel came along. They tried various schemes, but the bottom line is that all sufficiently complex mathematical systems are incomplete – they can express Gödel sentences that are true but can't be proved.'

'Yes, and we only know the sentences are true because we can see the problem from a higher level.' She decided it was now safe to put the crucial question to him. 'Does that mean that all sufficiently complex mathematical systems have a higher level since they all have Gödel sentences?'

David froze. He was looking at her, not seeing her, his full attention focused on his inner world. It was a little disconcerting – to stare back at him almost seemed like intruding. She lowered her eyes, waiting for him to speak.

Then, with a flick of his head, he was animated again, shaking himself out of his reverie – a dog shaking a wet coat.

'Sorry, Lucy, you started off a terrific train of thought there! That was the jackpot question!' He looked at her directly, realisation dawning in his eyes, his face serious. 'But you already knew that, didn't you?'

106

Lucy glanced down at her cup. 'I was thinking about it this morning, and I did wonder if there might be a kind of Inverse Gödel Theorem that proves the existence of a higher-level system. I guess there isn't?' He had got hold of the idea, and now he had to run with it, take ownership of it.

'No, there isn't – at least, not until this moment!' He paused. 'Lucy, I don't know if I should be asking this of you – it's probably unfair on you and wrong of me – but do you think you could possibly have time to work on this with me?'

This was totally unexpected! Her first reaction was exultation that he should want to spend more time with her. Closely following that was her perennial concern about setting aside enough time to keep up with her work, and then relief that she had not alienated him by suggesting the idea of an Inverse Gödel Theorem, all of which thoughts were abruptly displaced by the realisation that she would not measure up to his expectations.

She started to put her worry into words. 'David, I don't know what to say. I'm really flattered – I'm immensely flattered – that you should think I could be of any use to you—'

David raised his hand gently from the table, stopping her in mid-sentence. 'Lucy, Lucy,' he said almost tenderly, 'I know you haven't spent time studying the formal mathematics of all this, but I have, and that's why we'd make – we *do* make – such a great team! Knowing the syntax – knowing the mathematical grammar – is a bit like being a technician: that's my bit. What you bring to the team is your happy knack of asking the right questions – the revealing questions, the ones that get to the root of the problem. Even if you never had another new thought about this, your idea of an Inverse Gödel Theorem makes the whole thing possible in the first place. Please say you'll join me!'

Lucy was melted by his words. 'Well, after that, how could I refuse!' She gave him a broad grin, sealing the deal.

'Brilliant! Thank you, Lucy. Look, I imagine you're worrying about the time commitment' (*he was perceptive*) 'but I promise you, I wouldn't be so foolish – so selfish, even – as to jeopardise your studies. We can do this sensibly.' He rose out of his seat. 'Do you fancy a refill?'

Lucy said she would and she watched him go over to the counter for another two cappuccinos, looking even taller than his six-two, towering over the two dumpy coffee ladies. He returned with the two cups and placed two foil-wrapped chocolate caramel wafers on the table beside them.

'I thought we could do with something to dampen down the caffeine infusion,' he said. 'Look, here's how I think it could work. I'll work on the formal maths and translate it into English so we can talk about what I've been doing and what we should do next. Even just discussing the work with someone else is going to stop me from going up blind alleys and making mistakes. How would you feel about meeting at weekends – something like we've been doing today? That way, it need only amount to a few hours each week – what do you think?'

'Won't it take up too much of your own time, though?' she asked.

'I've invested a lot of my time already. Anyway, my PhD work hasn't suffered – I'm still on track. What I'm doing for my PhD is fairly mechanical.'

'Then I think that's a great idea,' said Lucy, and meant it.

CHAPTER 8

Saturday 23 October

The following Saturday, David arrived a quarter of an hour early at the Coffee House. He had been even earlier at the Maidenhead Café on the previous weekend in order to secure a table – he hadn't wanted anything to go wrong on their first date (he was now daring to call it that in his mind). The Coffee House had been his suggestion for this, their next meeting, not because he knew it particularly well – his knowledge of the interiors of any of the cafés in the town was scant – but because the display in its small window had always struck him as intriguing when he passed by, with its large range of coffee pots, teapots and other china tastefully arranged to attract the collector.

He had not expected the shop to be so full, and a brief survey revealed no vacant seats. He positioned himself rather awkwardly beside the counter, half-turned to scan the area for any signs of patrons preparing to leave, trying not to catch their eyes directly. He had been looking forward to this afternoon for the whole week, indeed, ever since he had left Lucy after their meeting. He had spent the latter part of his time with her in the Maidenhead Café showing her some of the mathematical syntax that she had asked him to explain. He had, of course, seen her in the Friday tutorial, but he had had no pretext for asking her to stay behind, although she had flashed him a warm smile at the end of the session, telling him quietly as she passed him, so that only he could hear, that she looked forward to their meeting the following day.

The sound of a chair being dragged back alerted him to a male student-type in the process of standing up at the table nearest to the window and shrugging on his anorak. The youth looked over to him for the first time, sending a small shock of recognition through David's brain – it was the student whom he had seen whispering to Lucy in his tutorial two weeks previously. For a wild moment, David thought that he must be here to meet Lucy but then dismissed the idea as ridiculous. The student came over to David.

'Hi,' he said, 'you waiting for a seat?'

'Oh, hello, yes – sorry, I should know your name?'

'No reason you should – it's George,' he said, shaking hands, 'and I know you're David Lane, of course. Please take my seat, quickly, before it's nabbed.'

'Thank you, George,' said David, noting that the other seat at the table was occupied by a young woman, 'but I won't, if you don't mind – I'm going to be joined by someone else shortly.'

'Oh, that's OK – she's leaving with me,' said George as the young woman rummaged in her bag and then stood up to go.

David walked smartly over to the table, draped his heavy anorak over the seat George had occupied, sat down on the other, facing the entrance, and gave a cursory wave through the shelves of china to the parting couple as they passed the window. David was elated by what had just happened. Either George and Lucy had broken up or else he had been mistaken about them in the first place. Then again, George could be two-timing Lucy, but he hadn't looked furtive. Of course, Lucy might still have another boyfriend, but now David had cause to hope, whereas before, in his mind, at least, had been the certainty of George.

He told himself Lucy wouldn't be the kind of person who would change boyfriends lightly. She was a serious student, not one to waste her time, by her own admission. Of course, that could weigh against himself in her mental balance of boyfriend time against study time. On the other hand, she didn't consider him as a boyfriend – he realised

110

that, to her, he would be a combination of collaborator and mentor for an idea – a project – that seemed to fascinate her as much as himself. His motivation for suggesting the semi-formal collaboration had been mixed, of course. In fact, he had taken himself by surprise by asking her – it was uncharacteristically bold of him. But he had been caught up in the excitement of the moment: her idea of an Inverse Gödel Theorem was stunning. And he was conscious of something else quite wonderful. If anyone else had come up with the idea in what he had come to regard as his territory, he would have been ungraciously jealous, perhaps even dismissive. But with Lucy, he had felt stimulated, almost proud of her. It was the unselfishness in him that was striking, his genuine pleasure in her ability to stand back and see the big picture, which was how she was able to ask the right questions. He wondered whether, after talking to her on only three occasions, he could be falling in love. He didn't know – he had no experience.

As if on cue, the door opened and she came in, heart-stoppingly beautiful in her curvy grey coat. *Oh to be the cute fur collar that cuddled round her neck!* She spotted him and her smile lit up the café. He rose out of his seat and had to check himself from completing the welcome with a kiss on the cheek. Instead, recovering, he removed his parka from her chair and they both sat down.

Lucy said she would get the coffee because it was an opportunity to sample from the vast range of varieties of beans they had in the shop: she had become interested in coffee since she had been given a coffee maker by her aunt.

'How has your week been?' David enquired, when they had both sugared and milked their drinks.

'I went to a concert on Tuesday night at the Elder Hall,' she said brightly, stirring her cup, quite unaware of the turmoil of emotions that this casual comment was to arouse in David.

111

His first reaction was surprise that she had had the time to go to such a function on a weekday evening, followed immediately by a pang of regret that he could not hope to be included in her social life, at the same time recognising that there were no grounds for him even entertaining such a hope. He wondered if she had gone with anyone to the concert. 'What was the concert?' he asked as lightly as he could.

'It was the University Symphony Orchestra. They were playing Albinoni's oboe concerto in D minor followed by Beethoven's seventh,' she said, looking out of the window, wistfully, David thought. 'Friends of mine in the hall, Gillian and Sarah, they persuaded me to take a break and go with them.' She looked back at David. 'They had a point,' she conceded. 'I didn't join any societies in Freshers' Week like the others did. I sometimes think I'm the only one in the hall who doesn't go to some sort of society or club or something regularly.'

'I have a theory about that,' said David, who had cheered up on hearing that she had gone to the concert with two girls. 'How many of your friends who are going out regularly are doing physics or maths?'

Lucy smiled. 'Yes, I had the same idea. If it's true, I wonder if it's the intensity of the subject that stops us getting involved with all the activities or whether we're self-selecting geeks?'

'A bit of both, I imagine. Did you enjoy the concert, then?'

'More than I can say.' She became serious. 'Actually, I wept in the first bit,' she added.

Perplexed, David remained silent.

'You see, they played the adagio from Albinoni's oboe concerto in D minor at my Dad's funeral, and it's a lovely piece; it was one of his favourites...' Her voice took on a husky edge and she stopped.

David resisted an urge to lean across the table and take her hand. 'Oh, Lucy, I'm so sorry,' he said softly. 'Was it recently?'

'February. He passed away in February. He was only forty-nine. A massive coronary. He was a doctor, a GP. Ironic, isn't it? He didn't spot the signs in himself. He collapsed in the surgery, but they couldn't bring him round. All those doctors there and he still died.'

'And your mother?'

'She was devastated, of course. But she's a nurse – not in Dad's practice, she works in the hospital – and she's quite level-headed: she's come to terms with it, I think. Sometimes I would hear her crying in the night, though.'

They were silent for a bit, David wishing he could comfort her, trying to think of something to say that wouldn't sound clumsy. Eventually, he asked: 'Do you have any sisters or brothers?'

'No, I'm an only child. A spoilt brat!' She smiled. 'I think maybe that made it harder for me – having nobody my own age to share it with.'

'Oh! I'm an only child, too!' He paused, then continued gently, 'Your father died right in the run-up to your A levels – that must have been an extra burden for you.'

'Not really, to be honest. Dad's death crowded everything else out – nothing else came close. Nothing else seemed particularly important for months. I took the A levels like I was in a daze, like a mechanical robot. I was detached from the worries that everybody else seemed to have; I was just watching myself go through the motions. Anyway, it turned out all right in the end.' She opened out her arms, palms upwards, *look-at-me-now!*

'How have you been coping?'

'Oh, you know, I have my moments…' her voice trailed away again. She seemed to give herself a mental shake. 'Anyway,' she said, more brightly, 'how about you – are both your parents still alive?'

113

David paused. 'Actually, no, they're not. My mother died when I was an undergraduate, and my father died a year ago.'

Lucy's hand flew to her mouth. 'Oh God, I'm sorry! Here's me going on about myself, and you, you're an—'

David smiled. 'No, don't worry. I wasn't particularly close to my father, if I'm honest. I miss my mother though. Anyway, it helped that I had been away from home for some time when they died – I was here at university. I wasn't with them every day, if you see what I mean. Right now, Lucy, I'm feeling a lot sadder for you than I am for myself!'

'Gosh, there's no need to,' she said. 'Still, I'm touched that you care.'

David would like to have said something about them caring about each other, but didn't. Instead, he took a sip of coffee.

'Did you find it hard, not being religious?' she asked.

'You mean coping with their deaths? Well, to tell you the truth, yes, I suppose it might have been a comfort to have believed in a god, but there it is.'

'Please don't take this the wrong way,' said Lucy, 'because I mean it kindly, but don't you find your view of the universe a little bleak? I mean, didn't you even say that self-awareness was really just a neural network making a model of itself? You're such a warm person to have such cold thoughts!'

David laughed and blushed at the same time. 'I think self-awareness is a wonderful phenomenon, but it's not miraculous, not supernatural.'

'You don't think we have a soul, then?'

'Look, Lucy, I would feel a bit awkward debating whether we have a soul or not after what we've just been discussing. It would feel, I don't know – disrespectful, if you see what I mean.'

114

'That's sensitive of you, David, thank you. But you don't need to feel you're walking on eggshells with this. I'm quite robust, you know. Actually, I'd be interested to hear what you think.'

'Well, all right,' said David, doubtfully. 'Where to begin?' He thought for a moment. 'OK, first, would you accept that physiological processes like vision and hearing don't depend on there being a soul?'

'How do you mean?'

'Well, for instance, if surgeons implant electrodes on the visual cortex, they can evoke visual sensations by activating these electrodes to stimulate neurons in the visual cortex. So at that level, at least, there is no need for a soul to explain the mechanism of vision.'

'No, I suppose not,' Lucy agreed.

'Right. So now imagine that an engineer invents a semiconductor neuron that has the same functionality as a wet neuron – a biological one. So this semiconductor neuron has dendrites to collect signals from other neurons and it sums the signals and if the sum is high enough it generates pulses along its axon to other neurons. And its threshold levels for firing and summing and so on which would have been controlled by the changing chemical environment in the brain are instead controlled by environmental voltage levels derived from other, similar semiconductor neurons. Are you with me so far?'

'...ish,' said Lucy. 'My biology was never strong. But I get the principle – you could substitute artificial neurons for biological ones in the brain.'

'That's it. Now suppose that the entire visual cortex in your brain is replaced by these semiconductor neurons. Do you think you would be aware of any difference?'

'No, I don't suppose so,' said Lucy, cautiously.

'OK. Now suppose I extend the replacement programme in your brain and systematically replace all of your neurons with these semiconductor neurons, so that you still process

115

the messages coming in from the eyes, ears and the rest of the senses, of course. Let's say I do this gradually and it takes a month to complete the project. Do you think you will be aware of the change at any stage in the work?'

'Hmm. I grant you that it's difficult to see just what would make me aware of these gradual replacements of my neurons, but, at the end of the day, the thing that is me isn't the neurons, it's the pattern of activity that they support, isn't it?'

'Well, exactly!' David beamed. 'That's the point I was leading up to. It doesn't matter *how* the pattern is generated: as long as the substitute neurons are identical in function, there will be no perceptible difference in the mind of the person whose neurons are being replaced. We could even take it to extremes and replace each neuron with a simulated one in a binary computer that had connections running into the brain to link up with where the original neurons took input from the senses and connected with the muscles and so on. So the processing could be done on a computer in the corner of the room rather than in the brain. Of course, you would still think you were seeing, feeling and hearing everything with your body.'

'But the pattern is abstract, though, isn't it?' Lucy asked, her voice lacking conviction.

'Granted, the pattern is complex, but running it on a binary computer – even one with a vastly parallel architecture – means that the pattern can, in principle, be reduced to numbers. In a funny way, it's a bit like Gödel-numbering the brain.'

'Wow, that sure came out of left field!' David found the Americanism on Lucy's lips unaccountably fetching. 'So what's Gödel numbering?' she asked.

'Sorry, it was one of the tricks that Gödel used to construct his theorem. He reduced all mathematical statements to numbers using a simple code. It made it easier to handle the statements that way. But the real analogy with

Gödel, of course, is that both Gödel's sentence and self-awareness are self-referential.'

'Oh, yes,' said Lucy. 'Tell me more about how you get to self-awareness.'

'Well, as I said, I don't see anything mysterious about self-awareness,' said David, resting his elbows on the table and unconsciously steepling his fingers, lecturing mode. 'For a start, all creatures need to interact with their environment. They need to locate food and move away from predators and other dangers. Generally, they do this using neural networks, which may be quite primitive in some cases. These networks take input from their senses, such as taste, scent, temperature, touch, sound and sight and send output to their mouth, jaws, legs and so on. The patterns that the surrounding environment creates in these networks are models of that environment. Even the lowly frog has a model in its brain – if a fly comes within vision, its tongue comes out and zap! – the fly is captured.'

Then David recalled a lecture he had attended. 'In fact,' he said, 'you can cut the optic nerve of a frog, then rotate its eyeball upside-down' (Lucy mouthed a silent *Ugh!*) 'and the cut nerves will re-grow to find their original endings. And, of course, after the operation, the when a fly flits by overhead, the frog's tongue darts down instead of up! Anyway, some higher animals have a more complicated interaction with their environment. There are videos of crows that drop nuts onto pedestrian crossings so that passing traffic can crack open the nut. The advantage of using a pedestrian crossing is that when the traffic stops, the crow can nip out in safety to collect the kernel!'

'Now you're kidding me!'

'I know how it sounds, but no, apparently it's true.'

'Well, it's surprising to think how intelligent birds can be,' Lucy observed. 'It makes Hitchcock's film just that little bit more creepy. Actually, I must tell you later about something that happened with a bird today. But are you saying the crows are self-aware?'

'No, I doubt if they have much awareness of self, if any at all,' said David, 'but that's not my point. The crows clearly have a good working model of their environment, and that includes the behaviour of traffic, which could be a potential predator but in this case acts in the role of nutcracker. Now if you take chimpanzees, their models of the environment include themselves. A chimp in a zoo in Sweden used to collect stones together and hide them before the zoo opened and then later he would use them as missiles to attack visitors. So, when he was collecting the stones, he must have had an image in his mind of himself throwing them at people. In effect, he had a model of himself.'

'Yes, I see,' said Lucy, 'a system containing a model of itself is self-referential, just like Gödel's sentence, and the point you're making is that the chimpanzee must have been self-aware. But doesn't the same apply to the crows, then – didn't their environmental model include themselves eating the cracked nut?'

'Maybe not,' said David. 'Maybe the birds see only the nut in their model but not themselves, just as a cheetah sees only the antelope as potential food. When it pursues the antelope, it isn't necessarily thinking of itself doing that – it just *does* it. The crows probably didn't sit down and plan the whole thing from scratch – seeing themselves up in the telegraph wires dropping the nuts onto the road below. Very likely it was behaviour learned from the first time a crow dropped a nut accidentally into the path of an oncoming vehicle, and improvements would have been added incrementally over time.'

He paused to take a sip of coffee, Lucy following suit, swallowed and resumed: 'Going back to the chimps for a moment – what we don't know is whether the chimp's image included what the visitors might think of his behaviour and – here's a thought – whether the chimp

118

thought about what the visitors might think the chimp would be thinking when throwing the stones!'

'Yes,' said Lucy, 'that would presumably be an even higher level of self-awareness – when you think about what others think you're thinking!'

'And human beings clearly operate on at least that level, because that's exactly what we're doing right now!'

'Yes, I must admit I'd not thought of there being degrees of self-awareness before. I suppose, if I had thought about it at all, I'd just have said that an animal is either self-aware or it's not.'

'You know,' said David, 'it's interesting talking about this out loud, because I've just seen a *formal* connection between the Gödel sentence and self-awareness.'

'How do you mean?'

'Well, remember how I said that an English version of the Gödel sentence, call it *G*, would be something like *G cannot be proved*? Well, in that sentence, instead of just saying *G*, I could substitute the meaning of *G*.'

'Let's see,' said Lucy, taking up the thought, 'that would mean it would read "*G cannot be proved* cannot be proved". Is that what you mean?'

'Yes, exactly! And you can carry that on forever – " '*G cannot be proved* cannot be proved' cannot be proved" and so on. Well, imagine that *G* is the innermost doll in a set of nested Russian dolls – so it's the smallest doll. Then we place that doll inside the second doll, the next smallest doll, and say that everything inside the second doll cannot be proved. So the two nested dolls would be the real, three-dimensional analogue of *G cannot be proved*. Then we place these in the third-smallest doll and say that everything inside that third-smallest doll cannot be proved either. We carry on like that, and we have a visual representation of the meaning of *G* that can go on forever like the nested sentences. But notice that each doll, apart from the innermost one, contains within itself a model of itself.

119

That's the formal connection I'm talking about. The nest of Russian dolls is self-similar. Of course, something having a model of itself within itself doesn't automatically make it self-aware, or the Russian dolls would be conscious! But it is a necessary condition for self-awareness – it's self-referential, self-similar.'

'So, what else do you need for self-awareness?'

'Well, I would say that the model needs to be current, continually updated with input from the environment, and also with input generated from within the model itself, so that different scenarios involving the environment can be tested, played out, before implementing them in reality.'

Lucy smiled. 'Nevertheless, David, you haven't convinced me yet that we don't have a soul.'

'Oh, I'd forgotten about that! Surely the fact that you can, in principle, at least, describe self-awareness mathematically means that you don't need a supernatural thing like a soul?'

'What it means to me is that it's the pattern – OK, the mathematics, if you like – that is the essence of the soul rather than the body that houses it.'

'I agree with you about the importance of the pattern rather than the body,' said David, 'but I'm thinking of self-awareness when I say that, not the soul.' He felt uncomfortable again, discussing this, but added, 'I think you need to define the soul.'

Lucy thought for a moment. 'I guess what I mean by the soul is, yes, the self-awareness, if you like, that inner narrative that you have with yourself, but also the fact that this pattern will be preserved – somehow – when the body dies.'

'Which brings us back to the question of a god, doesn't it?' said David, smiling. 'If there were an all-powerful god – which I don't believe there is – but if there were, then, sure, I could imagine the god looking down on us and deciding to preserve self-awareness patterns.'

It suddenly struck David that they had not even started to discuss the Inverse Gödel Theorem. By arranging the meeting on the pretext of their project, he felt more relaxed than he would have done, had this simply been a date. In consequence, perversely, it seemed to him as though they had been granted licence to extend the subject of their conversation well beyond the original brief without him having to worry about how he was measuring up in Lucy's beautiful eyes.

Almost as though Lucy were reading his thoughts, she said: 'Talking of God looking down, wasn't that effectively how you said Gödel's theorem is derived – you look from outside the system and see that the Gödel sentence is true, even though it can't be proved true in the system?'

'Yes, that's right. A Gödel's eye view, you might say!'

'Ouch! So if we were to develop an Inverse Gödel Theorem, somehow we have to find a way to talk about Gödel's sentence being true in the higher system. So that's why you wanted to avoid using the idea of *truth* in the Maidenhead Café.'

David's skin tingled to hear her speak of *we*. 'Yes, you're right. I've been thinking about the problem myself since last week. Let me try out something with you. Just a moment.' He rose from his seat, went over to the counter, and returned with a tray and two empty cups. He set the tray on the table and positioned the empty cups on opposite sides of the tray, one on Lucy's side and one on his own.

'Right,' he said. 'This tray represents system X. You've still got the cup containing statements that can be proved in system X and I still have the one with the statements that can't be proved in system X. OK so far?'

'Yes, just like before, as you said.'

'But now there's a difference. The table is a higher system. Let's say it's system W. It contains the tray – system X – as a subsystem. But it also contains its own list of statements – this time, they're statements in system W.

So your coffee cup on the table contains statements that can be proved in system W this time, and mine on the table contains statements that cannot be proved in system W. Still with me?'

'Yes,' said Lucy, 'but I've a question, though. Does my cup of provable statements in system W also contain a copy of the statements that are in my cup on the tray – the list of statements that are provable in system X, I mean?'

'Ah, good point! Yes, if a statement is provable in system X, in the tray, it has to be provable in the higher system W, on the table – I'm obviously assuming that the higher system contains all the axioms and rules of the lower system along with some extra ones in addition. Of course, these extra axioms and rules mean that there will probably be statements in your cup on the table – in system W – that can be proved in system W that can't be proved in system X – in the tray. So your cup on the table will be fuller than your cup on the tray.'

'And the same is true of *your* cup on the table – the list of unprovable statements in system W?' she asked.

'Ah, that's different! If a statement is not provable in system X, in the tray, it just might be provable in the higher system W, on the table, because of the extra rules and axioms. So my cup on the table could actually contain fewer unprovable statements than my cup on the tray. Although, of course,' he added, 'there may be some statements that you can express in system W that you can't in system X, just because of the additional axioms and rules, and so there may be some additional statements that are unprovable in the W system which simply don't appear in system X.'

'All right, I've got that, I think,' said Lucy. 'So how do you deal with Gödel's sentence, G, now that you've got a higher level?'

'OK, remember that whether we tried putting either G or the negation of G in your cup – on the tray in system X – we got a contradiction? Well, now that we've got the higher system *W* – the table – the resolution is to put *both G and its negation* into *my* cup in system *X* on the tray.'

'Whoa! Hold on!' said Lucy, raising her hands, palms outward, pushing back, eyes closed below a frown of concentration. 'Let me think about that.' She opened her eyes and looked at David. 'Is that allowed – a statement and its negation both in the same cup?'

'Well, it certainly wouldn't be allowed in your cup,' said David, 'because that would mean that you could both prove a statement and its negation, like proving two plus two is four and also five. But now that we have the higher system, *W*, you'll see in a moment that it's not necessarily wrong to put them both into my cup in system *X* on the tray. Remember that the full statement of *G* is *G cannot be proved – in system X*. He held up his finger for emphasis. 'But it may very well be provable *in system W*, the table. If that's the case, we can put *G* both in my cup on the tray and in your cup on the table.'

'Ah yes, I see now,' said Lucy, 'because we know, of course, that *G can* be proved in the higher system – that's the whole point of Gödel's theorem, isn't it? But what about the negation of *G* being in your cup on the tray? Do we also put that in a corresponding cup in system *W* on the table?'

'What do *you* think?'

'Em – well, if *G* is in my cup on the table, I guess the rules would say that the *negation* of *G* should go in *your* cup on the table?'

'Exactly!' cried David, causing a couple of heads at nearby tables to turn round and Lucy to grin broadly. He continued more quietly: 'What that says is that the negation of *G – G can be proved in system X –* is quite impossible, certainly in system *X*, and, of course, in any system, including *W*, on the table.'

'So there's a rule here, isn't there?' asked Lucy.

'Yes. The rule is that sometimes you'll get a statement that, try as you might, you can't prove in the system and yet you can't prove its negation either. So they *both* go into the list of unprovable statements in that system. When that happens, you know something odd is going on, and so you look in a higher system and – lo! – you see that the statement is in the list of provable statements in that higher system and its negation is quite correctly in the list of unprovable statements in that higher system.'

'Of course,' said Lucy, 'as you said in the West Gate Bar, the Citizens of the Tray, in system X, cannot prove that they cannot prove the sentence G! *We* know that's true of course ... Oh no!' she cupped a hand over her smile: 'I'm not allowed to talk about *truth*, am I? Anyway, I guess what we're saying is that the Citizens of system X can keep trying to prove that G cannot be proved but they'll never succeed. This constant failure may lead them to suspect that G is genuinely undecidable, but they can't prove it. There must be a very good reason for their constant failure, almost a law.'

David nodded his approval. 'Very good. And, of course, that reason can't be revealed in system X, because that would be a proof. The proof has to be outside system X – the higher-level system.'

Abruptly, Lucy exclaimed: 'David, look at the table! Look what we've done! Forget the milk and sugar – just look at the table!' Her eyes were bright, her cheeks flushed.

David looked down. 'Sorry, what do you mean?'

She was excited, and a little breathless. 'Well, what's on the table? Two cups – and a tray containing two cups! The table contains a model of itself!'

'Hah! So it does!' David felt his own excitement rising now. 'And, of course, system X will contain simpler systems, so the tray should contain a smaller tray which also has two small cups – and an even smaller tray and so on!'

Lucy laughed. 'Self-referential, self-similar, just like the Russian dolls!'

*

Half an hour later, they stood at the summit of a path that edged along the low cliffs defining the north-east boundary of the town, gazing out towards a broad horizon where an overcast sky met a greyer sea, stippled with white-capped waves whipped up by a strengthening breeze.

'They call this *Sunset Strip*,' said David, 'because you can watch the setting sun from here – not today, obviously – but you must also get a spectacular view of it rising over the sea in the spring and summer.'

'You've never seen it rise in all your time here?'

'Sadly not. My body clock doesn't know about early mornings!' He would dearly have loved to suggest making an exception if she would accompany him to watch the sunrise one morning, but recognised the pipe dream for what it was.

Both had felt the need for a change of scene after their rather intense mental work-out, David, in particular, wanting to let the new idea of a hierarchy of self-similar systems settle, mature, become metabolized into his conceptual landscape, creating a firmer foundation for further construction when they proceeded to the next phase of developing their Inverse Gödel Theorem.

'It's interesting,' he said, 'how changing perspective, literally, like we're doing now, just looking out to sea from the cliffs, can also help you change your perspective in thinking about an idea. You know, when I was at school, I'd climb to the top of the hill near my home and watch the sunset. It opened your mind – anything was possible!' He savoured the illicit thrill, the bitter-sweet self-torment in allowing Lucy such close proximity to this innermost of his secrets – how he would fantasize about his dream-girl on that hill.

Lucy stood by his side, taking in the vista as she replied: 'I know what you mean about looking down on the world from on high – my best friend and I used to sneak into the foyer of a multi-storey hotel that overlooked the docks and we'd climb to the top of the stairwell to get a panoramic view of the ships and Southampton Water all the way down to the Isle of Wight. The view was breathtaking. We were only thirteen or fourteen and if anybody had noticed us they'd just have assumed our parents were guests at the hotel.'

David pictured Lucy at fourteen years old and wished he could have met her at that age. Then, with a shock, he realised that he would have been about twenty then and immediately dismissed the thought as he tried once more to forget about the disparity in their ages. 'So do you come from Southampton, then?' he asked.

'Yes, I was born and brought up there. When Dad passed away, I thought about changing my choice of universities to stay nearer home, but Mum said I should come here as I'd set my heart on it. What about you? Where is your home?'

'In the west of Scotland. Fort William's the nearest town. But my home is in a clearing in the woods.'

'Do you still have the house?'

David nodded. 'Yes, and I don't intend to sell it. Too much tied up in it – I've been going back to it regularly to make sure it's kept going. It's not too far, provided I keep working here. I'm hoping to get a post-doc – that's a post-doctoral fellowship – here in the department when I finish my PhD.'

'Are post-docs difficult to get?'

'With any luck, I should get one. You really have to have done a post-doc before you can get a permanent lectureship – that's what I'm aiming for.'

'Do you think it's possible that what we're working on now could form the basis of your post-doc?' Lucy asked.

'Hmm – I've been wondering about that myself. I've even thought of a name for the field – *metamathematical physics*! But I'd need to have written a paper or two to establish my credentials in the subject.'

'Do you have enough, now, to write about the Inverse Gödel Theorem?'

He thought about that for a moment. 'Good question. Shall we sit down?' he said, indicating several benches by the side of the path. They chose one facing north-east and sat side-by-side, watching the seagulls wheeling as they crested the cliff on the updraft of the easterly breeze. 'You know, quite a few students have lost their lives here, falling over the cliff edge.'

Lucy was sceptical. 'Surely not! There's a railing, and look, there's a warning notice just over there.'

'It's true, I'm afraid. Mostly it happens in the dark, in the early hours – they think they're taking a shortcut to the beach, maybe to a beach party, I don't know. The tragedy is, the same thing happens time and again. It's only long-term students like myself who have been here for years who see the pattern recurring.'

Lucy looked down, saddened.

'I'm sorry, I didn't mean to be so gloomy,' he said, contrite. 'Think of it as a safety alert in case you're ever out here at night-time. Anyway, you were asking if we had enough for a paper on the Inverse Gödel Theorem,' he said, brightening up, and he proceeded to summarise their discussion, counting off the points on the fingers of his left hand. 'One – we know that we can always construct a Gödel sentence in a sufficiently complex system. Two – however, the inhabitants of this system can never know whether any *particular* sentence is a Gödel sentence – only that there must exist at least one such sentence. Three – the construction of the Gödel sentence makes it undecidable in the system – it cannot be proved nor can its negation be proved – and the inhabitants may even be lucky enough to choose the Gödel sentence by accident. But, however hard they try, they cannot *prove* that the sentence they have felicitously chosen is unprovable in their system. Four – the thing that prevents them from proving this, however – the reason for their continued failure – is not available to them within their system, otherwise that reason itself would form the proof they sought. Five – so all they know for sure is that there has to be a Gödel sentence within their system and that there will be something preventing its identification which must reside outside of their system. The *something* that is preventing them from identifying the Gödel sentence is a proof in the higher system that the Gödel sentence cannot be proved in their own system. That's the essence of the Inverse Gödel Theorem and it seems to me that when you apply it to any sufficiently complex mathematical system, it will point to a higher system in which the Gödel sentence *can* be proved.'

'Right,' Lucy said with a grin, 'now for the big one – does that apply to the mathematical system that underpins our own universe – does that prove that there is a higher universe?'

David smiled. 'Well, I admit I did have that in the back of my mind by trying to present the argument from the point of view of the inhabitants of the lower system. In the case of the universe, I guess we could call the higher system a meta-universe.'

'I don't mean to be a wet blanket, David,' said Lucy, 'but even accepting the existence of a higher mathematical system, why does that automatically mean there is a higher *universe*? I mean, couldn't a higher mathematical system be described in our own universe?'

'Ah, that brings us full circle,' said David. 'You remember how I was telling you in the West Gate Bar that I spent ages trying to reduce the universe to mathematics?'

Lucy nodded.

'Well, I wasn't just trying to write down the laws that are obeyed by the physical systems of the universe – I was actually trying to show that all physical systems *are* just mathematics – that *everything* is just mathematics!'

'Oh!' Lucy exclaimed, wide eyed. 'I didn't realise that's what you meant back there. That was a subtlety that went completely over my head!'

'You thought I meant' – David cast around for a suitable example – 'like describing how gases behave using the mathematics of thermodynamics?'

'Well, yes, I suppose. Isn't that what you meant, then?'

'No, quite the opposite, in fact. That would be mathematics *describing* a physical system – a gas. I was looking for the mathematics that *is* gas!'

'How do you mean?'

'Well, for example, think of the constituents of the gas – individual molecules made of atoms that are in turn made of quarks and leptons, and maybe these will turn out to be strings, we'll have to see… But, ultimately, we get to a level where all of the characteristics of these constituents are completely defined by just a few basic mathematical relationships, or patterns, if you like. That's not to say that the state of all the elements of the gas at any given time is simple. You'd need a horrendously big book just to write it down. Even once you'd described the protons and electrons and all the other constituents of even just one molecule, you'd have to add a description of its motion, and then you'd have to do the same for all the other molecules in the gas. And the motion of each molecule is changing all the time because of the changing distances between it and all the constituents of every other molecule in the gas and not forgetting the container and the universe beyond, as well. But the equations themselves are relatively simple. It's like evolving a complex Mandelbrot picture from a simple root equation.'

'So where does that get us?'

'What I'm saying is that, if you follow it far enough, everything in the universe boils down to a set of simple mathematical relationships. So, rather than mathematical relationships *describing* the motion of a planet, it would be more accurate to say that the moving planet *is* a set of mathematical relationships. Sure, the universe is complex, but that complexity can emerge from very simple equations, very simple relationships.'

'I'm not convinced, David. Surely, ultimately, mathematics in a vacuum doesn't do anything – it needs something to work upon – a substrate if you like?'

David waited for a hand-holding couple to amble past their bench before replying, oddly embarrassed at the prospect of being overheard, as though a deep philosophical discussion was not a convention to be expected between a couple sitting together, and which, therefore, in a curious inversion of the normal social rules, should be kept private.

'I know what you mean,' he said, 'but remember this – at rock bottom, there are no hidden properties. The electron has spin, and it interacts through electromagnetic, gravitational and weak interactions and with the Higgs field as well, of course. That's it. Other properties follow from *how* it interacts, like charge and mass. If any of these interactions turns out not to be the last word, then whatever supplants them will be just as simply described. There's nothing else to say about the electron, no hidden substrate. It is just these simple mathematical relationships. It's the same with the other particles, the protons and neutrons and the various fields. The reality of the things we perceive – the bench we're sitting on, seagulls, ocean waves – this reality results from the agglomeration, the hugely complex interaction, that emerges from a few fundamental mathematical relationships. This isn't a belief – it's rather a recognition.'

'So are you saying that reality is an illusion? Surely mathematics is something that ultimately you can describe,' she waved her hands, searching for an illustration, 'with equations on a piece of paper? The paper is just as real as the mathematics!'

'Exactly. The paper is as real as the fact that two plus three is five – in fact, it is real *because* two plus three is five. It is the reality of the mathematics that underpins the elementary particles of the paper, as well as the elementary particles of the neurons in your brain which perceive the paper, that makes the paper real.'

Another couple strolled by, but this time David carried on.

'What I'm saying is that a model of external reality seems to work well – we all agree that we can see a piece of paper with writing on it. But remember what we were saying about you and your awareness of your surroundings being a pattern of electrical activity in the brain. The atoms of the paper are ultimately pure mathematical relationships, which, incidentally, also account for the photons reflected from it onto your retina, eventually causing flurries of activity in certain neurons of your brain that are particularly sensitive to some characteristics of paper with ink marks on it. So you *recognise* paper with equations on it. Your neurons, the paper, the photons – they're all mathematics. But the feeling you get on recognising the paper is just as real as the paper itself!'

'So you would say that we're a bit like the people in the computer-simulated world in *The Thirteenth Floor*!' said Lucy.

'Not a bad analogy, if you're careful not to carry it too far. What *The Thirteenth Floor* has over *The Matrix* is that complete characters are simulated in the computer world, and they simply don't know they're simulations – everything they perceive around them makes up their reality.'

'Yes, and in *The Matrix*, the people exist independently of the computer – it's just that their brains are being fed with a computerised simulation. I like the way that *The*

Thirteenth Floor had the idea of people in a computer simulation devising a computer to create a computer-simulated world. A bit like our Russian-doll self-referential universe!'

'Yes, but don't look too closely,' David warned her again, good-naturedly. 'The difference between our universe and the scenario in the movie is that, to create our universe, there is no physical computer crunching away at the mathematics. There is no infrastructure, no' – he searched for an adjective – 'no *metacosmic* blackboard on which the mathematics is worked out – unless, maybe, you count the mathematics itself.'

He had another thought. 'Ironically, when we write down mathematics on a piece of paper, that's actually a case of the universe modelling mathematics – not the other way round! The equations on the paper aren't the mathematics – they're symbols we just use to *represent* mathematics by activating symbols in our brain – neurological patterns – that we have taught ourselves to associate with mathematical concepts. So if you see the sign for *plus* on the paper, you have a concept in your brain of taking two quantities – quantities which correspond to other concepts in your brain, incidentally – and then putting them together in a special way to make a different quantity. But that's not mathematics – it's merely a *model* of mathematics. In fact, since you *are* mathematics, you could say that when you look at an equation on a piece of paper, what you have is mathematics *modelling* mathematics!'

'In reality!' added Lucy.

'Sorry, I do go on, don't I? Shall we walk down to the pier?'

They resumed their stroll eastwards along the path, which now hugged the high wall surrounding the grounds of the ruined 12th century cathedral and would eventually take them down to the harbour.

'Is this idea of the universe *being* mathematics your own idea?' Lucy asked him.

'Well, I thought it was at first, but I soon found it has a respectable following among physicists, and quite a few philosophers and mathematicians as well. In fact, some of them argue that it helps to explain why the fundamental constants in our universe have the values that we see.'

'What, you mean the speed of light and that sort of thing?'

'Yes, that's right. Why, out of all the values that the speed of light could have had, and the charge on the electron, and all the rest – why these particular values and not others?' He broke off. 'Look, see that hole in the wall?' They stopped opposite a shallow buttress protruding from the cathedral wall and he pointed to a narrow, vertical, two-foot-high opening in the buttress like an arrow-slit set at the height of a toddler. The interior of the hole was pitch black. 'If you put your hand through that after midnight, the White Lady who haunts the gravestones on the other side will grab your arm and not let go.'

'Now you really are kidding me!'

'So you'd do it then – you'd come out here after midnight and put your hand through the hole?'

'Absolutely not! Oh, I'm not afraid of ghosts – just nutters who hang around cathedral walls after midnight!' She smiled. 'Anyway, what do you mean by saying that the fundamental constants are explained by the universe being mathematics?'

'Ah, you're not going to like this! You start off by assuming there is no God!'

Lucy was sceptical. 'What has that got to do with it?'

'It's quite important, really. You see, the combination of the values of the fundamental constants in our universe is just right for stars to live long enough to generate atoms with higher atomic numbers than just hydrogen and helium, and to disseminate these atoms in supernovae explosions which can be incorporated into planets orbiting other stars and eventually for life to evolve on some of these planets. Not only that, but there are relatively few *other* combinations of values of the fundamental constants that will allow life. Some people say that proves that God must have arranged our universe to be like that – for life to evolve.'

'That seems like a better argument for God than many I've heard,' said Lucy.

'Yes, but there's an alternative explanation that doesn't depend on having to invoke a god. Many physicists take seriously the idea that there is a whole host of parallel universes – a multiverse – with each universe having slightly different combinations of values of fundamental constants. In most of these parallel universes, stars won't be formed, or they won't last long enough for life to evolve. Only in universes like ours will life forms evolve that can ask the question – why is the universe just right for our existence? But that's just like having a whole lot of planets in our solar system and on one of them, Earth, a life form evolves that says – why was the Earth made just right for our existence – not too warm and not too cold? Well, of course, there is nothing special about Earth in the scheme of things – it's just that the other planets are too cold or too hot for life like ours. And it's the same with all these other parallel universes. We couldn't have lived in the inhospitable ones.'

'OK, putting aside for the moment the question of where all these parallel universes are, where does the mathematics come into it?'

'Ah, that's the clever bit. Assume that a universe *is* mathematics, or, at least, the *mathematical expression* of simple relations corresponding to the fundamental constants and to the elementary particles and to the basic interactions between them. These relations will be embedded in a mathematical system that allows interactions to add

134

together, maybe, or to be inverse-square, and so on. Then each different parallel universe can be the expression of the same mathematics but with different starting values for the fundamental constants and so on. The mathematics then takes care of how each universe develops. One of these will be our own universe. But there's a more interesting refinement, although it's more controversial.'

Lucy raised her eyebrows in mock incredulity. 'You mean parallel universes are *not* controversial?'

They had just resumed their walk along the path, but, before David could reply, they were both spattered by large, stuttering raindrops borne on a sudden downdraft of wind, and they ran back to seek shelter in the corner between the buttress and the cathedral wall, the precipitation escalating rapidly into a fluent curtain of rain. The shallow buttress afforded only moderate protection and they flattened themselves with their backs to the wall, Lucy right in the corner and David so close to her that he could feel her shoulder next to his right biceps.

To David, it was unutterably romantic, watching the rain drenching down in sheets while they stood snug and dry in their little niche, just the two of them in an otherwise deserted landscape. He controlled a strong desire to extend a protective arm over her shoulders, berating himself for his adolescent urges, and continued with his explanation, doing his best to filter out a sudden gruffness that threatened to modulate his voice.

'As well as having parallel universes for a range of different values for the fundamental constants, you could also have a range of different mathematics.'

'I don't follow,' said Lucy, from somewhere below his right ear.

'Well, for instance, the mathematics for a universe might require two particles to interact as though there were an inverse-square force between them. But it could equally appear as inverse-cube or whatever. If we allow such a range of mathematics, many, probably most, universes would be unstable, but they would still be allowed, just as ours is. The thing is, the mathematical function that describes all of these possibilities will generally be much simpler than each separate possibility.'

'How do you mean?'

David risked a sideways glance. She was looking out towards the sheets of rain falling beyond the cliff edge, unaware of his silent appraisal. He was so close to the top of her head that he could easily have buried his face in her shiny brown hair, mesmerised at being so near to the essence of her thoughts – her being – just centimetres below the chestnut sheen.

He looked away again. 'It's easier to explain that with an analogy. Suppose that every universe is described by some positive integer, like a Gödel number, a code. Of course, because there are so many universes and because each one has a certain complexity, the code number of each universe will generally be very large, although not infinitely so. Anyway, you can specify all the possible code numbers with a very simple mathematical relation – you just start with the number one and keep adding one to the total. That would cover every possible universe. But for a *given* universe you would probably have to specify a very large number, because of the very large number of possible complex universes. It's the same with mathematics. The root mathematics may be comparatively simple, but any particular expression of it for a given set of universes can be quite complex.'

'OK, I think I get it now,' said Lucy. 'The idea is that, if you accept that mathematics *is* reality, then quite a simple set of mathematics leads to a whole host – a multiverse – of parallel universes, one of which has the same laws, like the inverse-square law of gravity, and the same mix of fundamental constants as ours, and so that allows life – us –

to evolve and ask the question – why is the universe just right for us?'

'Yes, and, paradoxically, the complexity of any one of these universes is apparently a lot higher than the fundamental mathematics that creates them.'

'So, if the multiverse exists, then it exists at a higher level than its individual component universes, including our own. It's what we called the meta-universe, isn't it? In that case, surely what we've done is not really new?' Lucy sounded disappointed.

'Well, in the first place,' said David, 'we derived the meta-universe from basic mathematical principles, whereas the multiverse was suggested as the result of alternative, parallel solutions to postulated quantum mechanical equations of the universe, helped along because it explained our universe having the right fundamental constants for life to evolve. In the second place, your Inverse Gödel Theorem *is* new, at least as far as I can tell. I'll do another literature search and then write a draft paper on it and you can see what you think of it. It shouldn't take me very long because I can build on previous work in the literature that I've already got.'

Lucy was clearly touched by the compliment he had paid her in acknowledging her contribution, her voice sounding diffident as she protested 'It's not *my* Inverse Gödel Theorem – we worked it out together!'

'Don't underestimate what you've done, Lucy! I know within myself that I would never have stood back far enough to see that I was looking at the problem the wrong way round. But arriving at the multiverse from the Inverse-Gödel angle would be an independent route to the multiverse – and maybe even more credible than the one based upon the fundamental constants being different in parallel universes. So, yes, I would say it's new.'

The rain had begun to ease off, the sky brightening by the minute.

David added: 'But I think we should avoid discussing the meta-universe as a possible consequence of the Inverse Gödel Theorem in the paper – that would be too provocative for many physicists and mathematicians. What we need to do is to get the paper accepted first to give us some street-credibility before we start making extraordinary claims about a meta-universe. Oh, and by the way, you're first author!'

'Oh no, I couldn't be! I shouldn't even be on it – you're the one who's writing it!'

'Yes, and you're the one who came up with the idea, and you're the one who keeps coming up with new angles. I'd have had absolutely nothing to write about if it hadn't been for you.'

'Then I should go second on the paper, if you really want to include me,' she offered.

'OK, since it's so difficult to decide this, let's just do it alphabetically. Oh, look, *Darling* comes before *Lane* – that's settled, then, you're first.'

Lucy tried to protest further but she couldn't get the words out for laughing. David, exultant to see her so joyful, wondered if she had noticed that their initials were the reverse of each other's. He had spotted the small coincidence the first weekend they had met, and had filled a page of A4 paper, doodling with her name and initials like an infatuated teenager. Nevertheless, while he was well aware that many physicists in his position would place their own name first in similar circumstances, he was convinced that the correct thing to do was to put hers first, as the one who had come up with the idea.

'I'm sorry, David,' said Lucy, becoming serious, 'I really mean this. I absolutely cannot have my name first. Apart from anything else, suppose it got published and somebody asked me something about the paper. I wouldn't even be able to understand their question, most likely, let alone answer it.'

'All right,' David relented, 'I can see you're determined about this. Agreed, then, me first and you second. But we both know where the inspiration for it came from.'

When the rain had eased off, they decided to leave the pier until their next meeting, and to get back to their respective homes before the next shower. David turned off at Spey Street, but stopped on impulse after a few yards to watch Lucy continue on her way to her residence. His heart did a somersault when she looked back, saw him standing there, waved and then walked onwards to New Hall.

CHAPTER 9

Saturday 23 October

Lucy's heart mirrored the acrobatics of David's when curiosity made her turn round to see if she could find out where his flat was and saw him just standing on the pavement, watching her. Flustered, she waved and then hurried on. For an instant she wondered if he had been about to call out to her, something he'd forgotten to tell her, perhaps, but a moment's reflection satisfied her that such an explanation was unlikely: otherwise, he would have gestured, shouted, even chased after her when he saw her looking back. No, however she viewed it, any plausible interpretation involved David being *interested* in her, at least to some degree.

These pleasurable musings occupied her mind her all the way to New Hall, so that, as she entered the vestibule, she realised that she could not recollect any part of her journey there after waving to David. Passing the coffee bar on the way to the lift, she was spotted by Gillian and Sarah who called out to her. She was tempted to carry on as though she had not heard them, to continue up to her room and indulge herself in thinking about that last encounter, but, on hearing her name, she had broken her stride – a giveaway – and, in any case, she was sufficiently elated about David that she was ready – hoping, even – to be teased about him, as she suspected might happen.

True to form, as she joined them, her friends greeted her with the news that she and David had been seen together that afternoon (*was there no privacy in this small university town?*). 'Emma saw you both sitting at the window of the

Coffee House in Whitefriars Garden,' said Gillian, making no attempt to keep the triumph out of her voice.

'Oh – I didn't see Emma, then.' Emma was a first-year friend in common to the group.

'That's because you couldn't drag your gaze away from David! We heard you were gazing into those limpid blue eyes of his all afternoon.' This from Sarah.

'That's utter nonsense, and you know it,' said Lucy, laughing despite herself. 'Anyway, who said his eyes are blue?'

'Why, what colour are they, then?'

'Well, they *are* blue, as it happens. But that was just a lucky guess!'

Sarah pounced. 'Ah, so you *do* know the colour of his eyes! QED!'

'Why, what are we trying to prove here?' Lucy asked.

'That you're serious about David, of course! Why else would you meet him for coffee on a Saturday afternoon?'

'As it happens, we only met up today because he's helping me with a project,' said Lucy. This was almost true – she knew it would sound absurd to tell them that it was the other way round, that in fact he had asked her to help *him*. Moreover, while she enjoyed being teased about the growing feelings she secretly harboured for David, she dared not admit to anything but an academic relationship, even if Gillian and Sarah suspected there were more to it, because, while they would swear to keep it to themselves, the story would get out, maybe even reach David's ears, which would be disastrous.

'He must be kept very busy, then,' observed Gillian with light irony, 'meeting all his students individually to help them with their projects!'

'Yes, I suppose he must,' said Lucy innocently, not rising to it.

'It's the old Chinese proverb,' Sarah pronounced. 'If someone saves a life, then they're responsible for it ever after. David saved you from choking to death, so now you belong to him!'

Lucy quite liked the sound of that. She wondered if David had heard the proverb.

'And you have him for tutorials every week,' said Gillian. 'Does it feel strange in the tutorials, knowing him – ah – more personally, as you do?'

'Yes, I suppose it does, rather. I try not to draw attention to myself too much and so I don't really join in a lot.' Indeed, in both of the tutorials since she had started meeting with David, she had continued to sit in the back row, not particularly engaging with the discussion, but nevertheless satisfying herself that she seemed to be measuring up to the best of the students who were taking a more active part in the proceedings. Her contribution to the tunnel-through-the-Earth discussion had been an exception.

When they could see they were going to get no further admissions from Lucy, Gillian and Sarah allowed the conversation to turn to their own stories about the lectures they had attended.

'We had a really thought-provoking session on linguistics,' said Gillian. 'The lecturer asked the students to call out the number *two* in as many different languages as they knew.'

'OK,' said Lucy. '*Two, deux, zvei, dos.*'

'See anything in common?'

'Well, they're all monosyllabic. They all begin with a consonant.'

'Anything special about the consonant?' Gillian asked.

'Oh, I see, they're not the same consonant, but they all start off by putting your tongue on the roof of your mouth and forcing the air out.'

'Very good! In fact, nearly all the European languages have similar-sounding words for *two*. But then it gets more interesting. It turns out that in Hindi, it's *do*, in Urdu, it's *douh* and in Sanskrit, it's *dwo*. And there are lots of other basic, common words, like *mother* and *father* and so on that sound similar when you say them in all these different languages.'

'Oh, that *is* interesting. So English and the other European languages share a common root with these Indian ones?'

'Yes, it looks that way,' said Gillian. 'The lecturer showed how they derive from a common language, an ancient language, called Proto-Indo-European, that must have been the root of most of the current European and Northern Indian languages. They reckon there was a tribe wandering around Siberia, southern Russia, around seven thousand years ago, and that's where it all kicked off. Of course, there would have been other languages pre-dating Proto-Indo-European going back a long way, but we have no real clues about them.'

'I like that. It's like discovering the stories behind simple nursery rhymes, like the plague interpretation of *Ring a Ring o' Roses*. It's almost as though the people who lived hundreds or thousands of years ago are speaking to you directly, through the language of today, if you can only decode it.'

'I don't think the plague explanation is supposed to be creditable any more,' said Sarah.

'What decided the words in the first place?' Lucy wondered aloud. 'Was Proto-Indo-European supposed to be as complex then as our languages are today?'

'Apparently not,' said Gillian. 'And not all languages are equally complex, even today.'

'Although you can get some surprising concepts in some languages,' said Sarah. 'You have to ask why they needed a word for them in the first place. The lecturer gave the example of the word *rawa-dawa* in the Mundari language. They speak it in parts of India and Nepal and Bangladesh. It means – wait for it – the feeling that you get when you suddenly realise that you can get away with doing

143

something really awful because there's nobody around to see you doing it!'

'What – all that in one word?' said Lucy dubiously.

'I'm not joking,' said Gillian, who now had a fit of the giggles. 'Have you ever done anything to give you rawa-dawa?'

They discussed various scenarios interrupted by frequent hoots of laughter, causing several students to look towards their table.

'But it raises an interesting point,' said Lucy, sobering up. 'When you have a simple word for a complex idea, you can think more complex thoughts. You don't spend time and energy on working out what the concept is, and you can concentrate instead on manipulating the concept itself. For instance,' she said, her voice threatening to relapse, 'you could say that the boy's rawa-dawa deflated like a punctured balloon when his mother came in.'

Amid the guffaws, Lucy persisted. 'You see – if you had to discuss the situation without that word, it just wouldn't be funny. So the word adds complexity and helps you to think. Maybe language is a tool that helps us to model our environment in our minds?' … *and model ourselves?* she wondered.

'But surely language is primarily a way to communicate?' said Gillian.

'Yes, I'm sure you're right. That must be how it arose. But once it was out there – once people were able to verbalise their thoughts, I would guess that language would actually contribute to the mental picture we have of ourselves. Words are bouncing around my head all the time when I'm alone – doesn't that happen with you?'

'She's hearing voices in her head,' said Sarah mischievously to Gillian, tapping her temple with her forefinger, and the three of them burst out laughing again.

'I think language is like a mathematical system,' Lucy persisted, when the noise had subsided. 'It's like a system of equations that express concepts succinctly. You can use the equations to work out how things interact with each other, and you know in the back of your mind that you can always translate the equations back into their basic meaning. But, without the equations, you'd simply be unable to handle the sophisticated relationships between concepts.'

'So a dictionary would be the equivalent of you translating the equations back into what they mean?' said Gillian.

'Yes, that's right.' An idea occurred to her. 'Has it ever struck you that there's no such thing as a perfect dictionary?'

'What do you mean?'

'Well, while a dictionary might be self-consistent, in that every word in it will be defined in terms of words elsewhere in the dictionary, if the dictionary was your only source of information, and it didn't have any pictures or anything like that in it – only words – then you'd never understand anything!'

They looked at her blankly, waiting for her to explain.

'Well, think about it. Say it was a Chinese dictionary – the words would be meaningless squiggles to you, and, while you might see a pattern emerging between the squiggles that defined the other squiggles, you'd need something outside of the dictionary to start you off.'

'Oh, yes, I see what you mean, now' said Gillian. 'You'd at least need some pictures in it. And complex notions like *rawa-dawa* would need a lot of pictures before you got the idea!'

'You know,' said Lucy, reflectively, 'by coincidence, that concept is related to the project that David and I have been working on.'

'What – *rawa-dawa*?' said Sarah.

'No, I'm being serious. You know how you were saying the dictionary was like the system of equations? Well, just as the dictionary can't completely define itself, neither can

145

any given mathematical system. That's kind of what we've been doing.'

They looked at her as if to say *pull the other one.*

CHAPTER 10

Monday 25 October

In a reprise of his visit two weeks previously, Jeremy Wilgoss opened the door to the laboratory just enough to peer round it before venturing fully into the room, reminding David of a tortoise tentatively extending its head from its shell to scan the horizon before making its next move. David had been down visiting Mike, as he often did on Mondays, and the discussion had soon turned to David's supervisor.

'So, how are you getting along with Wilgoss's experiment, then?' David had asked.

'I've made some progress,' said Mike, beckoning David to follow him. 'There's something odd about the project I don't get, though,' he added, as they inspected the apparatus set out on the optical bench.

'What do you mean?'

'Well, you know he said that the Abraham-Minkowski stress tensor question was a century old and hadn't been resolved yet?'

David nodded.

'Well, in a way he was right – this *has* been controversial for that long, but the puzzle has really been resolved now. Abraham and Minkowski each calculated a different tensor for the momentum of light in a transparent material – that's more your field than mine. Anyway, various physicists tried all sorts of experiments for decades and didn't really come to an agreement.'

'But now they have?'

'Yes, more or less. And the theoretical side has largely been settled, too. Essentially, it turns out that both Abraham and Minkowski and all the experimenters were right – it just depended upon whether you were talking about the momentum of the photon itself or whether you included the momentum of the material. It's notoriously difficult to separate out all of the effects as you can imagine.'

'So, are you saying Wilgoss's experiment isn't worth doing?'

Mike looked glum. 'I'm afraid that would be my guess. It's difficult to see what this experiment will yield that hasn't already been published. I suppose our experiment could be seen as confirmatory, and it would be quite nice if I could get my name on the paper.'

'Are you going to ask him about it, then? I mean the fact that the work seems to have been done already?'

'No way! You know how difficult Wilgoss can be.'

David had been about to agree when they were interrupted by someone at the door, who turned out to be Wilgoss himself.

After the due pleasantries, Mike proceeded to show Wilgoss the progress he had made in setting up the experiment. Despite Mike's reservations about the relevance of the project, he had clearly taken great care in constructing the optical circuit, and explained to Wilgoss where and how he had had to improvise over the original design that Wilgoss had sketched out for him. Although he had known Mike for years, David had not, until now, realised just how good an experimentalist he was. As he listened to Mike's detailed description, he became increasingly impressed with the improvisation and ingenuity that had gone into this experiment.

Wilgoss, however, seemed indifferent to the effort that this must have cost Mike, so much so, in fact, that David, annoyed with Wilgoss, found himself asking Mike about the experimental arrangement in an attempt to compensate for Wilgoss's apparent lack of interest.

Then, unexpectedly, Wilgoss asked if Mike had found that the experiment was disrupting his PhD work.

'No, it's OK,' he answered. 'I tend to split the time between both.'

'So you won't be running out of helium, then, for your plasma vessel.'

This was an odd, even absurd, *non sequitur*, but, to his credit, Mike didn't miss a beat. 'Not at all. We have *L* size cylinders of high-purity helium in there' – he nodded to the gas store – 'that's the largest size there is. Anyway, the plasma only requires a moderate flow.'

'And how do you control the flow?' Wilgoss asked.

'There's what's called a pressure regulator valve fixed to the top of the cylinder. I control it that way.'

'I'd be interested to see it – would you mind?'

Mike darted a look to David, who could only shrug, as astonished by such an odd request as his friend was, and opened the door to the gas store. Although there was a window, it was blackened by a well fitting blind, in order to maintain the integrity of the light seals of the optics laboratory when the door opened. Mike switched on the light and pointed to the gauges at the top of one of four man-sized cylinders chained to the wall. 'This dial shows the pressure of helium in the cylinder – look, it's nearly full, 200 bar. This dial shows the pressure that's going into the plasma vessel, and this wheel adjusts that pressure.'

'And these green cylinders – what are these?'

'Oh, they're argon. I mix a little argon with the helium: it increases the efficiency of electron production.'

'You have two of each gas, then – presumably duplicating for backup?'

'More or less,' said Mike. 'There's always at least one full cylinder of each gas so we don't run out of either helium or argon.'

149

'What's the volume of gas in the full cylinders, do you think?' asked Wilgoss.

'About ten cubic metres at atmospheric pressure,' said Mike, checking the dial.

'Ah, so no chance of it running out during the experiment then, I imagine,' said Wilgoss.

'None at all.'

'Very well, thank you. It seems to me that you should be able to complete this without further input from me, then.'

David could not help smiling to Mike from behind Wilgoss's back, but Mike played it straight. 'No, I'll let you know when it's finished,' he said.

Satisfied, Wilgoss crossed the laboratory and made for the door, turning at the last moment to address David.

'Tell me, how is your thesis going?' he asked, his voice raised to carry across the room over the background thrum of the vacuum pumps.

David was stunned. Over the past two years, he had talked with Wilgoss no more than a dozen times, and not once had Wilgoss expressed any interest in the progress of his thesis.

'Yeah, fine. It's fine, coming along fine,' he said, matching the volume of Wilgoss's question. (*Was that all he could think of saying?* But he hadn't been prepared for such a question, not in a million years.)

'And have you started to write up yet? Final year, you know!'

'Uh, sure, Jeremy, I've been thinking about that, too. I'm just at the stage of putting something together, in fact.'

'Excellent!' beamed Wilgoss. 'Perhaps you could come by my office, say, later this morning, and let me have a copy of what you've done so far?'

'Uh, sure. Well, actually, Jeremy, I really haven't anything on paper that I could show you. Not right now, that is.'

Wilgoss continued to stand and look at David, the radiance of the smile perplexingly undiminished by the news. And so it was in that long moment of suspended animation, crying out for a happy word to be dropped into it to break the spell and release him, that David felt compelled to fill the silence and utter the words that would later cause him so much heartache and distress.

'Actually, I would have had something earlier, but I've been working on a paper in parallel,' he said.

Wilgoss was clearly taken aback. 'A paper, you say!'

Immediately David realised his mistake. The last thing Wilgoss needed was to feel left out, particularly by one of his students apparently trying to gain the kudos of publication without his knowledge.

'I am delighted that you have been putting your time to good use,' said Wilgoss, recovering quickly. 'Perhaps you would indulge me by letting me see the manuscript before we publish?'

'Uh, yes, of course, Jeremy. I was going to bring it to you next week. It should be readable by then.'

'Splendid. You may know that I am on the editorial board of the *International Journal of Natural Philosophy*. Of course, we shall submit it to that journal and I can, ah, ease its passage through the editorial process.'

'Actually, there are two of us writing the paper,' said David, somewhat embarrassed that Mike, apparently busying himself with making adjustments to optical components on the granite bench, was nevertheless a witness to the exchange, however unwilling.

'Good. I shall leave you to apprise him of our conversation, then. And, since this will be your very first paper, you can tell him also that I shall put my name last. I shall see you in my office next week,' and, with that, Wilgoss departed.

As David stared at the closing door he heard a snort and Mike looked up from the bench grinning broadly. 'Coming for coffee?' he said.

CHAPTER 11

Saturday 30 October

Lucy thought she had never seen the North Sea look so beautiful as she headed eastwards from Sunset Strip and down towards the pier. True, she had never in fact seen the North Sea at all before coming here to university, but she was enchanted by how calm it was and by the way the morning sun was reproduced in a myriad of sparkling wavelets on the flat, blue water across the bay. It was only ten o'clock, very early on a Saturday by student standards, which had the benefit that there was practically nobody around the harbour area to break the spell. They had arranged to meet in the morning rather than later because David wanted to leave for his old family home after lunch to check on the integrity of the building and be back by nightfall on the Sunday.

Reaching the harbour-side, she looked out along the pier, a solid, stone-and-concrete jetty stretching for some eight hundred feet eastwards into the sea, forming the main breakwater for the pleasure boats and the small fishing fleet that still used the harbour. The ebbing tide had exposed a sandbank on one side of the harbour, but there was still enough water on the other side to float some of the smaller vessels that hugged the base of the harbour wall and to nudge some of the larger ones lurching drunkenly on the bottom, focusing the sun in dancing caustics onto the sides of their brightly painted hulls.

David had said he would meet her on the pier itself – if it had been raining the fall-back plan had been to meet at a café *en route* – but there was no sign of him: for once, Lucy had arrived first. She decided to stroll out towards the end of the pier from where she would be rewarded with a panoramic view of the town when she retraced her steps.

As she set off along the pier she allowed herself to recognise, and even accept, the feeling of contentment that had begun to enfold her. She felt more at peace now than she had ever been since her Dad had passed away, and she was pleased to find that, rather than feeling guilty about her happiness, she could almost sense his pride and blessing in what she had achieved since his passing: her entry into an ancient and prestigious university, her lucky choice of residence, the fact that she seemed able to manage her time and cope fairly easily with the work despite her initial reservations and, of course, meeting and working with David.

Indeed, as she now acknowledged, her association with David seemed to suffuse all of the other components of her contentment in a kind of holographic principle. She was well aware that this was a fragile construction, because it was founded on a relationship that was essentially undefined between them. For her part, whenever her mind was not directly occupied with a problem or task, she would find her thoughts homing in on him. For her, the topography of the week was a saw-tooth – from Sunday onwards the anticipation of seeing him ramped up to a climax, culminating on Friday afternoon and Saturday, dropping to the baseline once more on the following day.

For David's part, though, she could only speculate and hope. He had certainly given her no overt signal that he was interested in her in any capacity beyond that of scientific collaborator, which was fair enough, as that had been the deal at the outset. However, there had been no mention of his having a partner – indeed, she was not at all sure that he had even said he shared his flat with anybody. Then there had been the prize piece of evidence – his standing watching her as she left him to go home.

154

The bottom line, though, was that, against his experience and maturity, she was too young and, yes, innocent, to risk taking the initiative. She knew that, if she asked Gillian and Sarah for their advice, they would both urge her to make the first move. However, their experience had presumably been honed on boys of their own age, which would not necessarily extrapolate to her own situation. She could hear her Mum saying that she shouldn't rush it, but that love would blossom in its own good time if it was meant to.

She had just concluded that the best advice would be her Mum's (albeit, of course, that it was actually her own advice to herself, repeating what she believed her Mum would have said to her) when she noticed a pair of trousered legs protruding from the wall ahead, approximately half-way towards the end of the pier. In addition to sheltering the harbour from incoming waves, the breakwater also served to protect its visitors from northerly winds by means of a wall running the full length of the pier on its north side. This wall was up to a couple of metres broad in places, serving as a high-level return path for the more intrepid students who regularly ventured to the end of the breakwater. Cut into the wall, about half-way along, was a seat in an alcove, hiding the sitter from the view of pier-walkers, apart from the legs.

Lucy's pulse beat faster as she became convinced it was David. Sure enough, as she drew level with the alcove, David looked up from a printed paper he was reading and greeted her with a smile that lit up his face.

'Good morning, Lucy,' he said, patting the stone bench beside him. 'Didn't we choose a brilliant day?'

She nodded, sitting down. 'I've been amazed at how good the climate is here – totally unexpected.'

'I know, it's surprising, isn't it? Last Saturday in October! Oh, tomorrow it will only be nine o'clock when the sun's this high – don't forget to put your clock back tonight and get an extra hour in bed.'

155

Lucy indicated the paper he had been studying. 'That's not it already, is it?'

'Well, it's a first attempt, at any rate,' he said, unfolding the paper back to the title page. 'Here, let me take you through it.'

They bent down to peruse the manuscript together, their heads nearly touching. Lucy read the title: *Undecidable propositions as evidence for self-similar meta-systems* and, underneath, their names – *three* names!

'I'm guessing you've got to the part where the name *Wilgoss* pops up?' said David.

She had, indeed. Unlike David, though, she was not at all exercised by the inclusion of the extra name: she simply assumed from it that students must have to submit papers for publication with the endorsement of their supervisors, signalled by the inclusion of the supervisor's name. However, when she checked this supposition with David, he blurted out an ironic 'Hah!' and went on to explain that supervisors generally had to do *some*thing to earn their place on the paper.

'Still,' he added, relenting somewhat, 'I suppose I can see a few advantages to Wilgoss being on it. Despite his reputation with the students, having him as a co-author will lend some weight in getting it published, especially with him being on the editorial board of the journal. And I checked the kind of stuff they publish in the journal – it's the *International Journal of Natural Philosophy* – they accept pure mathematics papers as well as fundamental physics papers, so, in fact, it's quite appropriate for us.'

For the next hour, as the climbing sun warmed and bathed their secluded, south-facing alcove in golden autumnal light, David went systematically through his manuscript with Lucy, explaining the meaning of the symbols and the logic so that the conclusion of the paper – that, for any conceivable, sufficiently complex mathematical system, its ability to express a Gödel sentence

156

implied the existence of a higher meta-system with the same basic structure – seemed satisfyingly inevitable. From time to time, Lucy would ask him to clarify a point, and occasionally he would find an error or a better way to express himself and would make a note in the margin joined by a long, stringy arrow pointing to where he wished to insert his correction or addition. At the end of the process, Lucy was pleasantly surprised to find that she had been able to follow the gist of just about everything in the paper.

'I feel very much the passenger,' she told David. 'Not only have you written the paper but you've had to explain it to me word for word.'

'My goodness, Lucy, you are so modest,' he said. 'This paper would never have existed if it hadn't been for your insight. I just provided the technical jargon to wrap it up in. And going over it like this has helped me to check it through and hopefully improve it. Is there anything more that you can think of to change or add?'

Lucy laughed. 'I would never presume— Actually,' she added in afterthought, 'there may be one thing you might want to think about.'

'Yes, of course,' David prompted her, looking quite eager to hear what she had to suggest.

Lucy was relieved to have something to contribute at this stage. 'Well, it's just that, after the introduction, which really sets the history and the context of Gödel's Incompleteness Theorem, the paper gets serious very quickly – it launches straight into the proof.'

'Hmm, yes, you're right, I see what you mean,' said David slowly, head down, scanning the first page again.

'Well, would it be OK to put in a paragraph just to give an overall idea of where the paper is going? I know that if I were reading it for the first time, I'd find it a useful guide – a kind of route map, you know?'

'Goodness, yes, you're absolutely right!' he cried, sitting straight up. 'In fact, even Gödel did that, in his original paper. It's called a proof-sketch.'

'A proof-sketch! Well, what I'd like to read in the proof-sketch would be more or less what you said on that bench on Sunset Strip last week. You know – we start by saying that a Gödel sentence can be stated in any sufficiently complex system, call it system X, but the sentence is designed in such a way that neither it nor its negation can be proved in system X. However, the reason for this cannot be demonstrated in system X because that very demonstration would constitute a proof of the Gödel sentence within system X. Nevertheless, that reason must exist or it *would* be possible to prove either the Gödel sentence or its negation in system X. Therefore, the reason, which is another term for the proof, must exist in a higher, meta-system that is not logically accessible to the lower system. In that meta-system, it is possible to prove the Gödel sentence for system X. This implies that there is a higher, meta-system.'

'Bravo!' said David, cheering her on. 'And we can add into this proof-sketch the bit about the meta-system being similar in structure to the original system. I mean, in both systems, statements may be categorized as *provable* or *unprovable*. All of the provable statements in the lower system will appear again in the meta-system, although, as we saw in the Coffee House, the unprovable statements in the lower system may not all be reproduced as unprovable in the meta-system. Also, in both systems, for every statement in the *provable* category there is a negation of the statement in the *unprovable* category, but if a statement is in the *unprovable* category, its negation could *also* be in the *unprovable* category – which would make it an undecidable proposition – and, if that is the case, then, in a sufficiently high meta-system, either the statement or its negation – but

158

not both, of course – will be in the *provable* category of that meta-system.'

Thanks to their coffee-table conversations, Lucy was gratified to find that she could more or less understand what he had just said. 'Yes, I think I follow you. I've just had a thought, though. What about the other aspect of the system and the meta-system being similar? We agreed that the meta-system contains the original system as a kind of subset, because all of the provable statements in the original system must also be provable in the meta-system.'

'Yes, you're quite right to point that out. We need to make it clear that each mathematical system is defined not only by its rules and axioms but also by all of its provable and unprovable statements. The whole system will be a subset of its meta-system. It's in the equations, but it would be good to spell it out at the beginning – the meta-system contains the original system as a subset, and then that original system contains a lower-down, less powerful system as a subset of its own, and so on. Yes, we need to include that – these are all the things that go to make the hierarchy of systems self-similar.'

Lucy was delighted at being able to show David that she was not only able to grasp the gist of the argument but that she could contribute to the paper. Then another question struck her: 'Why couldn't we apply the Inverse Gödel Theorem to *any* system – wouldn't that prove there must be an infinite number of systems, all self-similar, nested like an infinite series of Russian dolls?

'No, I don't think so,' said David. 'If you take a sufficiently low-level system, then it won't be powerful enough to construct the self-referential Gödel sentence.'

'What about the other way, then? What about going upwards, in the direction of higher-level systems?'

'Well, in principle it must be possible to construct a Gödel sentence in the highest-level system we know – the mathematical system that generates our own universe – call

159

it the *universal mathematical system.* That would give us the undecidable proposition that would imply the existence of a higher-level, meta-universal mathematical system that generates the meta-universe. But that process may not continue upwards forever. For instance, mathematicians have suggested extending the natural numbers to what they call *generalized natural numbers.* These would not only include all the ordinary natural numbers like zero, one, two, three and so on, but also those that you'll never reach by simply counting, because they're infinite.'

'Sorry, I don't get it. Why does that stop you from going forever upwards in a ladder of meta-systems?'

'Ah, because, in a meta-system that included generalized natural numbers, you could add the negation of the Gödel sentence as an axiom, a fundamental truth.'

'*What?* I thought we ruled that out before? That would mean you could prove that something couldn't be proved – it's self-contradictory.'

'I know, and we were quite right to rule it out – *as long as we were only using natural numbers.* But the proof that this axiom would offer in the generalized-natural-number system would only be possible using the numbers beyond the natural ones – the *supernatural* ones, if you like. They've actually been called that, believe it or not. Anyway, the upshot is that Gödel's sentence would no longer be an undecidable proposition in this generalized-natural-number system. So there would be no need for even higher meta-systems, at least, not on account of Gödel sentences, because they would have lost their power to point to higher systems, not being undecidable propositions any longer.'

'That is truly weird.' She thought about this for a moment. 'So, does that mean that the ladder of meta-systems stops once you include generalized natural numbers?'

'No, not necessarily. There *may* be higher levels, but we just wouldn't have the Inverse Gödel Theorem telling us

160

that there have to be. So I think there may well be an ultimate high-level meta-system – call it the God System!'

They watched, mulling over what David had just said, as two men descended to the largest fishing boat in the harbour – the *Highland Queen*, Lucy recalled – in preparation for the incoming tide.

'Oh, there's another thing I've just thought of,' she continued. 'At the end of the paper, should we mention that we're working on another paper about how to apply the Inverse Gödel Theorem to our universe and that this may indicate a meta-universe?'

'You're right to raise the idea,' David agreed, 'and, in fact, it's quite a common ploy for authors to say what they hope to do next. It's a bit of a risk, of course, because it might give ideas to rivals, but it also stakes out the territory to some extent – the more scrupulous workers might steer clear of working on the idea themselves and it also means the authors can show they thought of the idea first if someone else happens to put the idea into print in the not-too-distant future.'

'So you'll add a bit saying that we are going to apply the Inverse Gödel Theorem to the universe, then?'

'Well, actually, in our particular case, I think it would be better not to – it might be counter-productive for us. You see, for one thing, the meta-universe is quite a radical idea – after all, it depends on accepting that our universe is pure mathematics – and anyway, the readership of this paper may not be the same lot that would even be interested in applying it to the universe. I think mentioning a meta-universe might lose us credibility in the eyes of the referees of the paper, which would mean it would likely be rejected, despite Wilgoss sponsoring it. I suggest we leave out the bit about meta-universes until the paper's hopefully published. In fact, best not to tell a soul.'

'Ah, I hadn't thought of it that way,' said Lucy. 'That sounds like the wisest thing to do.'

They paused to watch a young lad walk past holding a fishing rod and carrying a haversack on his back, heading outwards towards the end of the pier. Apart from themselves, he was the only person who had been on the pier since they arrived.

'Oh,' said Lucy, 'I've just remembered something I meant to ask you last week. When you're talking about parallel universes and the mathematics that generates them, what sort of pictures do you get in your mind?'

'Oh, now *there's* an interesting question! Well, to start with, the picture has to be a bit fluid. That's because there are umpteen different, competing pictures suggested by cosmologists, and all of them have some merit, but none of them come with hard evidence. For example, some people think that space is infinite and expanding from one big bang, and that, while we can only see in principle to the limits of our own universe – you could call it our local *Hubble volume* – there is an infinite number of such Hubble volumes in this infinite space, each expanding, and each of which would be a different universe – you could say they would be parallel universes.'

'Oh,' said Lucy. 'Actually, that's not what I'd envisaged as a parallel universe.'

'No, I know what you mean, but wait, there's more. Some cosmologists think that, instead of one big bang, there is an infinity of big bangs, each with its own spacetime. Others have suggested that parallel universes are separated from each other through extra spatial dimensions although only by a microscopic amount. In some of these models, universes exist that are identical to ours, although unimaginably far way in space, and in other models, these parallel universes are branching off from earlier universes all the time, so they are very close in some ways but entirely inaccessible in others.'

Lucy felt oddly deflated by all this. If there were so many models, they were clearly all speculative, or else one

162

of them would long ago have been selected above all of the rival ones. But presumably this also meant that any mental picture that David had of his meta-universe would only apply to *one* of these models, and, therefore, stood a good chance of being wrong when the correct model was finally identified?

She put this to David, who seemed reassuringly cheerful in the face of such doubts.

'It doesn't really matter which picture you like best,' he said, 'as long as you accept that, ultimately, these universes are pure mathematics. What we do in our second paper will apply to any of the models, and to any model proposed in the future. You see, the difference between what the cosmologists are doing and what we are doing is that the cosmologists are suggesting models of the beginning of the universe and maybe other, parallel universes that depend upon laws that we may be familiar with or maybe laws that are alien to our universe. But what *we* are doing is to ask the question – *what permits the laws to exist in the first place?*'

'OK, then maybe I'd understand it better if I could see the picture you have in your head when you think about the mathematical systems and the universes they generate.'

'You're quite right, Lucy, I should have made all this clearer. I suppose I'm just a little afraid of giving you wrong information, because any picture of this is inevitably going to be flawed. But, anyway, here goes.'

He paused, evidently deciding how best to describe what he considered to be the central idea. 'The universe – or a meta-universe for that matter – resulting from the expression of a mathematical system will consist entirely of provable statements made using the rules and axioms of that system. Are you OK with that?'

Lucy nodded.

'Well, remember we said that all the provable statements that a mathematical system generated were part of that mathematical system?'

163

'Yes'

'OK, then that must mean that a universe generated by a mathematical system is an *integral part* of that mathematical system. So I regard ourselves and our universe as part of the universal mathematical system.'

'Oh, I see, now. That's different from what I'd been imagining. I'd had a picture of the universe and its mathematical system as somehow separate – even though I see now that you were telling me differently all along.'

'I know, it's a lot to swallow in one gulp.'

Lucy had another thought. 'What do you think about the meta-universe – do you see that as separate from ours or are we somehow part of it, too?'

'I suppose the way I picture our universe and the meta-universe rather depends on the question I'm trying to answer. Take quantum physics, for instance. My guess is that quantum-mechanical phenomena that we see in our universe will arise quite logically from the axioms and rules of a sufficiently complex meta-universe. So, in that instance, I suppose I think of our universe as a flat, two-dimensional plane sandwiched within a three-dimensional meta-universe. A sphere – a meta-sphere – travelling in the meta-universe that crossed through the plane of our own, flat universe would be perceived by us as a circle that burst into existence and grew to a maximum size and then shrank and disappeared again. Well, in fact, of course, we would see it as a *sphere* that grew from a tiny seed out of nowhere and then collapsed and vanished. Anyway, it would be quite mysterious to us, but not to observers in the meta-universe. That's just a picture, of course, not meant to be taken literally, but I think that some of the weird stuff in quantum mechanics would equally look logical in the meta-universe. So, to answer your original question, yes, I see our universe as part of a meta-universe.'

Lucy thought about this. 'So are you saying that the universes generated by different hierarchies of universal mathematical systems and meta-universal mathematical systems are all in the same place?'

'In a way, yes, but rather than talk about them being in the same place, because then you're thinking of three dimensions, which not all universes will have – some will have more, some less – it might be better to say that each universe is a part of the one above it. I don't mean *above* in the positional sense, of course, but in the hierarchical sense. By the way, there may be many universes in one meta-universe – these will be parallel universes, and not part of each other. So I guess the ultimate picture I have in my mind would resemble a family tree – I have a vague notion of a single ancestor universe at the top of the tree with branches coming down from it – again, I'm talking hierarchically, not spatially – leading to successive generations of more and more possible parallel universes. Each generation is produced by its meta-universal mathematical system.'

Mention of the family tree suggested another analogy to Lucy. 'So our own universal mathematical system could be like DNA,' she said, the intonation making it a question, 'and each parallel universe generated by it would be a different expression of DNA?' In the distance, back on the mainland, she spotted a dog walking beside its owner along the harbour-side. 'So, if our own universe were a dog, say, then you'd find dog-DNA pervading everything in our universe, and if a parallel universe were a cat, you'd find cat-DNA all through it.'

'That's not a bad way of looking at it,' David agreed, 'because that's an important feature of our universe – or any universe. I mean the idea of the universal mathematical system pervading our universe – that's why mathematics is so successful in explaining features of the universe. In fact, your analogy may be better than you intended, because if you just look at a DNA molecule, you would never know that you could grow a dog from it! Well, the universal mathematical system must be just like that – you could never predict all the detailed complexity of our universe from just looking at what must be a relatively simple mathematical system.'

He paused, waiting for Lucy to come back, but she was still thinking about this and so he continued. 'And don't forget that this mathematics even generates the space and time for its universes, and the other dimensions, too. Exactly what kind of universes or even what structures are generated – one big bang, multiple big bangs, whatever – I couldn't tell you. But what I *can* tell you is that the basic mathematics must be capable of generating our own universe – because the reality is that we're here.'

'I'd like to come back to what you mean by *reality* in a moment,' said Lucy, 'but I just want to confirm my understanding – the Gödel sentence is not generated by the universe itself, but by the universal mathematical system?'

'Yes, that's right – using your analogy, you could think of the Gödel sentence as being generated by the DNA – the same universal mathematical system that generated the universe.'

'But we're still saying that the universe itself is mathematics?'

'Yes, definitely. The universe *is* mathematics – abstract relationships, like the properties of the electron and how the electron interacts with other mathematical entities like protons and photons. But these relationships have many apparently arbitrary characteristics – like the ratio of the proton mass to the electron mass being around two thousand. Well, that's because the relationships themselves, including these ratios and so on, are generated by the DNA – the universal mathematical system. But that system will also presumably generate an indefinite number of slightly different relationships and ratios – these will be different, parallel universes.'

166

Lucy caught him glancing surreptitiously at his watch, and remembered that he wanted to drive to his old home that afternoon. Covertly, she checked the time herself – nearly mid-day. The boy had reached the end of the pier and was setting up his fishing tackle.

David continued: 'I've just thought of another analogy that might help with the concept of thinking about the universe as purely mathematical – and where – if that's the right word – all of the parallel universes are. Think of any of these computer games where you are chasing something, or being chased, or say it's a flight simulator and you want to land a plane safely,' he said. 'Well, the space and time that it takes to land the plane doesn't exist, of course. It's purely mathematical. There isn't a long semiconductor chip that is actually the runway and a smaller one that is the plane. Yet it's all entirely consistent, and, if the computer model were good enough, we would even have a model of the pilot modelling all this in his brain.'

'Like the *Thirteenth Floor* again,' said Lucy.

'Yes, but, as I said, I shouldn't take the analogy too far. Anyway, you could have the computer running several simulations at once, using the same mathematics, the same program, in other words, but with different input parameters – maybe a different model of aircraft, say, or a night landing, perhaps. My point is, you don't ask where both airspaces are, physically. The computer mathematics is capable of an indefinite number of different scenarios or universes, and, in principle, they don't take up any space at all!'

Lucy was silent while she thought about this. David pointed to the end of the pier. 'Look, he's caught something, hasn't he?'

'Looks like it. That was quick! What kind of fish do they catch here, do you know?'

David admitted that he had never fished in the sea before, and suggested that they go along to find out. They

167

walked side-by-side to the end of the pier and started a conversation with the lad, who looked young enough to be at school. He was evidently delighted that they had seen him land the fish, which, he assured them, was a flounder, now laid out, flat and dead, on the pier beside his haversack. He dug out a metal tape measure from his bag, pulled on the tab and read off the length of the fish – about a foot.

'I should think about getting back,' said David, to Lucy's disappointment, although she had known he couldn't spend much longer with her. So they started back towards the mainland, a couple of seagulls wheeling round behind them, threatening to steal the boy's catch. This put Lucy in mind of the pigeon that had killed itself on her window pane, and she recounted the tale to David as they walked towards the shore.

'When I looked down at the bird, it made me think of God watching us,' she told him, 'and then I realised why your argument about God not being omniscient doesn't hold water.'

David turned and grinned. 'Why not?' he asked.

'Well, because, of course, God doesn't inhabit our universe. I mean, He's not bounded by it. After all, haven't we just shown the possibility of a God System? So He can know everything about our universe – He is indeed omniscient.'

'Lucy, I hope you realised that I had my tongue firmly in my cheek when I referred to the ultimate meta-system as the God System?'

She laughed. 'Don't worry, David, I was using the term in the same spirit, if that's the right word. But it's a handy name for the ultimate meta-system, don't you think?'

'Maybe it is, but there's nothing to say that there is a self-aware being in the God System, let alone one that would be interested in our particular universe. Anyway, I thought that part of believing in a god is accepting the god without proof of where it is or how it comes to be there or how it operates or anything?'

'Well, yes, that *is* the Christian view. "Blessed are they that have not seen, and yet have believed." That's what Jesus said when Thomas insisted on feeling Jesus' wounds before he could believe that Jesus had been resurrected. I won't deny that there needs to be a degree of mystery for Christianity to work.'

'That probably applies to all religions, for that matter,' said David. 'Well, you can still have your mystery.'

'How's that?'

'Well, I don't see how we can prove much about the God System. It seems that the Inverse Gödel Theorem indicates that such an ultimate system may well exist – and that's a tremendous claim, if you think about it – I don't think anyone has come anywhere near as close as we have to proving the existence of a higher universe, let alone the God System – but we can't prove what the mathematics of the God System would be, because our own universal mathematical system simply isn't powerful enough. In fact, it may be the ultimate mystery!' He stopped speaking for several steps, deep in thought, and Lucy kept quiet.

'One thing I think we *can* say,' he resumed, 'is that, in one sense, we can view the God System as a direct, single step from our own universal mathematical system. That's because, while there may well be a whole hierarchy of meta-systems above our own, we could view them as being concatenated into one single meta-system. In fact, we could go on adding to the hierarchy of meta-systems, one upon the other, until we came to the end – and my guess is that there *would* be an end, because, eventually, the Inverse Gödel Theorem will lose its power, as we said.'

'Why do you say that you can treat a whole hierarchy of meta-systems on top of our own as just one meta-system?' Lucy asked.

'Well, any theorems that are proved in a lower system must still hold in the higher system that contains the lower system as a subset. Actually, you can see that in the equations in our paper. It would be a bit like starting with the mathematical system that deals with the natural whole numbers – you know, zero, one, two and so on, and then finding that there is a higher system that deals with fractions, and then a still higher one that includes negative numbers, and then an extension of that into the irrational numbers like the square root of two, and an even higher system that includes imaginary numbers like the square root of minus one, and so on. Each system in the hierarchy will enfold the theorems of the lower-down systems, and add a few of its own, along with new axioms.'

'So, from the point of view of the highest system, all the subsystems below it are transparent?' said Lucy.

'That's right, because they're all contained within the highest system, along with their theorems. But, of course, it doesn't work the other way round – you can't see what's going on in a higher system from one lower down. You may guess, but there can be no proof.'

'As shown by Gödel's theorem'

'Yes, exactly.'

They had reached the mainland now and had started to walk along the harbour-side, past stacks of lobster pots and nets and thick, blue ropes used by the diminished fishing fleet that still provided lobster, crab and langoustines for the local community. All of the boats were now completely afloat, buoyed by the incoming tide over the past two hours. Lucy was suddenly struck by the disparity between the extremely abstract discussion that they had been having and the reality represented by this picture-postcard scene, illuminated by a surprisingly dazzling autumn sun, populated now by a few strolling tourists and students.

She voiced her thoughts to David. 'Thinking about the meta-universe is almost like being in a dream. When I'm

170

talking to you about it, or even just thinking about it to myself, the meta-universe seems logical and real enough – an almost inevitable result of the Inverse Gödel Theorem. But then when I come out of the dream and back into reality – to this,' she gestured across the harbour, 'then I think I must just have been dreaming before – how could the meta-universe – even the God System – be real?'

'Actually, Lucy, I know what you mean. I have the same kind of thoughts myself, sometimes.'

'Do you? So what do you do about it?'

'Well, I look down at my hands and remind myself that the atoms in my body were once part of a star that exploded, dispersing the atoms into space. As Carl Sagan put it so wonderfully, *we are star-stuff.* That reality is no less than the reality of – what – of this stone arch.' They had turned from the harbour now, and were heading under an ancient archway and up the road skirting the south wall of the cathedral.

'You certainly put that vividly!' said Lucy. 'I guess what you're saying is that it's a matter of perspective. It's easier to accept the reality of things that are nearer to us, and that are closer to our size and work on a timescale that is similar to our own.'

'That's right. In fact, you put it very well yourself, remember? You talked about people building chalets in the Alps and yet they may never stop to think about the momentous forces that those very same Alps are subject to, continually being pushed up by the collision of continental plates. That was because we see things on a human scale.'

Lucy was pleased and touched that he had remembered the conversation of their first meeting. And he had referred to Carl Sagan, one of her own favourite figures in science.

'Well, we can *demonstrate* the reality of the arch,' she said, 'but we can't demonstrate the God System, can we?'

'Well, maybe,' he said, slowly. 'It depends what you mean. I would say that the result of applying the Inverse

171

Gödel Theorem to the universal mathematical system is a compelling indication that there *is* a higher system, and maybe an ultimate system – what we've facetiously called the God System. What's going to be more difficult, though, will be to find a consequence of the God System for our universe.'

'You mean, can our knowledge of the God System be used to predict something that hasn't yet been observed in our universe?'

'Yes, precisely. It wouldn't *prove* the existence of the God System, of course, but it would be evidence in its favour.'

'What sort of prediction would people look for, do you think?'

'That's tricky. Part of the problem is that the existence of the God System is such a general result. It might be a factor in explaining phenomena that seem counter-intuitive, like much in quantum physics, but it's going to be difficult to come up with specific predictions. That can be our third paper,' he added with a laugh.

As they approached South Street, they worked out arrangements for Professor Oakhill's party. Although this would not be for another two weeks, David could not meet Lucy before then as he was to attend a conference, along with Oakhill among others, as it happened, on the weekend before the party. Since Oakhill lived out of town, they agreed that David would pick up Lucy in his car and drive the two of them to the party. Lucy left David half-way down South Street, as he was heading directly for his car, kept in a lock-up garage in Alexandra Place owing to lack of space at his flat in Spey Street, and she set off for lunch at New Hall.

CHAPTER 12

Saturday 13 November

Standing just inside the entrance to New Hall, trying to remain still or, at least, not to make any sudden movement that would trigger the infrared sensor controlling the large glazed sliding doors and allow antisocial levels of chilled air to penetrate the warmth of the foyer, Lucy continued to vacillate over her choice of clothes for Professor John Oakhill's party. In making the arrangements for David picking her up, she had asked him what people had worn there on previous occasions. David could not exactly recollect how they had been dressed (*how like a man*, she thought, warmly) but he had the impression that the postgraduate students, at least, had been in the usual jeans and T-shirt. When she had pressed him about his own plans, he denied even owning a pair of jeans, which had added weight to her choice of a skirt. Throughout the week, however, she had changed her mind several times, her fear of being over-dressed competing against her need not to appear disrespectful to her eminent host.

A pair of headlights swung into the large horseshoe driveway and the car stopped at the entrance – too late now to change her mind. The reflection of the brightly lit foyer in the glass doors made it impossible to see who it was, and so she walked forward, the doors parting to reveal David emerging from the far side of the car.

'Hi, David,' she called out to him. 'It's OK, don't get out – I'll get in myself,' and she opened the passenger door and slid inside, holding her gift bottle of red wine.

'You look nice,' said David as she clunked the door shut, 'perfect for the party.' She could have hugged him for that. He had said it slightly awkwardly, as though unused to paying women compliments, somehow adding to his sincerity and at once reassuring her.

Lucy's immediate impression of the inside of the vehicle was how luxurious it was. The seats were upholstered in ivory leather, the fascia was trimmed with a polished wood veneer, and, as she settled into the seat, she felt as though she were easing into a first-class seat on an airliner, or, on second thoughts, the flight deck of a modern fighter aircraft, a sensation reinforced as the courtesy lights dimmed to reveal a multitude of instrument lights, green, amber, blue, yellow, sprinkled like Christmas tree decorations on the dashboard, on the sound system, even on the steering wheel and the wide armrests on the doors. Handel's *Sarabande* was playing in the background.

'I can see why you don't leave this in the street,' she observed dryly, buckling her seat belt.

As they drew away from the kerb, smoothly, almost silently – she noticed that the car was automatic – David sounded sheepish. 'Yes, it is a bit over the top,' he said, punctuating the admission with an apologetic laugh. 'I tell myself that I need the comfort so I can drive the long miles back to the west, but the truth is, I just love it. I hope I'm not this profligate with anything else,' he added in mitigation.

'Oh, please, there's no need to defend it – it's a lovely car. It must be great to drive. How far is it to Professor Oakhill's?' she asked, changing the subject.

'Not far – it's about eight miles, I think. They have a very large house with a great view of the sea – but of course, you won't see the view tonight. It's actually two houses that were converted into one. A sad history, really, but it worked out in the end.'

'What do you mean?'

174

'Well, John Oakhill used to live in town, close to the university, in fact, and then his wife got breast cancer and died. Towards the end, he was visiting her in hospital and met the woman who eventually became his current wife. She happened to be visiting her husband who had something wrong with his heart, I think it was, and he died around the same time as John's wife. Anyway, the relationship blossomed, they married and, as far as I can gather, he sold his house and bought the one next to his new wife and they merged the two. They have five kids between them, so I guess they need the space!'

'A sad story with a happy ending! I wonder if the fact that they had both lost their spouses helped them, you know, to understand what the other was going through?' Lucy thought about her Mum – would she ever get married again? She would like to think so, for her sake, because, with Lucy away from home, she had nobody. Not that anyone would ever replace Dad, of course, but that wasn't the point.

She was startled by David's perceptive reply. 'How would you feel if your mother married again?' he asked her, gently.

'I was just thinking that very thought,' she said, the amazement showing in her voice. 'I hadn't considered it until just now, but, yes, I would be pleased for her – all alone, you know – if she felt able to marry again, I'd be really happy for her.'

The music system selected Pachelbel's *Canon*, the sweet violins resonant with the stately progress of the luxury ride, but the music was also a poignant, rather melancholic accompaniment to their conversation. Perhaps to lighten the mood, as they left the outskirts of the town and headed along a country-dark B-road, David began to tell Lucy about seeing Wilgoss on the Monday following their meeting to hand over the manuscript of their paper after incorporating all her suggestions. Wilgoss had wanted a paper copy as well as the electronic version and had

175

immediately coned down on the title page with the authors' names, presumably, said David, to check that his own had been included. Wilgoss had been taken aback to find that the third author was someone whose surname he didn't recognise. His disconcertation had increased with the discovery that the *L* indicated a female author, and a first-year undergraduate, to boot, not that there was anything wrong with a female author of course, it was just unexpected – *why?* – well, simply because he had never come across a female theoretical physicist, oh well, yes, except Lisa Randall, of course, and yes, Fotini Markopoulou, and, OK, Laura Mersini-Houghton and, yes, naturally Emmy Noether and, yes, indeed, there were plenty of examples, but a first-year undergraduate? The central idea in the paper was hers? At that point, Wilgoss had given way and they had moved on to discuss the contents of the work.

As anticipated, Wilgoss was unfamiliar with the field, and he had been content to listen to David going quickly through the pages, summarising the key points, answering his few queries and, in the end, accepting the paper without any changes, much to David's relief. Wilgoss would submit the paper through the normal electronic channels but would contact the editor separately, in his capacity as an editorial board member, to ensure that the work would have a smooth and rapid passage through the publication process.

As David pointed out, this was uncharacteristically generous of Wilgoss, unless it was simply a ploy to increase his publication record without expending any effort. He then went on to regale her with a repertoire of 'Wilgoss stories', some of which David had witnessed personally, many of which had become campus legends, all of which kept Lucy amazed and amused until they arrived at John Oakhill's, for which she was grateful to David, as she had grown increasingly unsettled during the week at the prospect of being by far the most academically junior guest at the party.

They parked in the street and David went round to the boot, emerging with a substantial cardboard box, garishly decorated in colours made even more lurid by the orange street lighting.

'Fireworks!' Lucy exclaimed.

'Yes, rockets. Didn't we say it was a firework party? It always is. I'm sorry you didn't realise – I just didn't think.'

'No, that's fine. It's great, in fact. Should I have brought fireworks instead of a bottle, though?'

'You didn't need to bring anything at all,' said David, 'but they'll be just as pleased with wine as fireworks.'

They were met at the door by Oakhill's wife, Mary, who thanked Lucy for the wine and ushered them through to the kitchen. 'You may want to keep your coats on until the show's over,' she said, and left them to help themselves to drinks and nibbles and then to diffuse, Brownian-style between the party-goers, towards the open back door.

Standing outside, on the paved terrace, about twenty people were chatting in groups and watching Oakhill and several others preparing the firework display in the darker recesses of the garden, a hissing, crackling bonfire illuminating their profiles like chiaroscuro figures in a Joseph Wright painting. David introduced Lucy to two postgraduates in theoretical physics while he made his way down to the pyrotechnicians to deliver the rockets.

Speaking to the students in her group, Lucy gathered that Oakhill generally invited the lecturers and post-doctoral fellows in theoretical physics only, presumably because there would not have been room for the whole department, especially as many of them brought along their children, too. However, in the case of the postgraduate students, the invitation extended to both theoreticians and experimentalists, which made for a good mix. She was gratified that her junior status seemed to make no difference

177

to them, although, despite their diplomatic constraint, they were predictably curious as to the circumstances of her being there. She felt cosily secure saying that she was David's friend.

As David returned to join her on the terrace, her smile of welcome froze on her face as, out of the corner of her vision, standing in a small group, his back to the garden and the bonfire, seemingly pontificating to his small audience, quite unaware of her presence, she caught sight of her humiliator, her nemesis, her co-author, Wilgoss. All the anxiety of the week, which David had helped to stem in the car, and which the informality and promised fun of a fireworks party had seemingly dispelled, now came flooding back. *Of course* he would have been invited, but it simply had not dawned on her all this time that, in coming to the party, she would meet the professor who had so publicly taunted her.

'What's wrong, Lucy?' David asked, the concern evident in his face and voice.

'It's Wilgoss,' she said urgently, quietly, so that he had to bend close to hear. 'Don't look, but he's over there at the far end of the patio.'

David straightened up and burst out laughing, causing a bolt of annoyance to surge through her, followed by a wave of burning embarrassment lest Wilgoss should hear David and look over to see what the joke was.

'Sorry, Lucy,' said David, immediately penitent, obviously seeing how important this was to her. 'It was just the way you said it – naturally he's here! Look, there's no way he'll ever remember speaking to you, I promise you. Not in the circumstances of that lecture. Anyway, he's not a great mixer – stay with the crowd and you won't even bump into him.'

They were interrupted by the sudden whoosh of a rocket snaking neck-craningly skywards, culminating in a bursting, expanding universe of blue-white stars and followed a moment later by a resounding thunderclap, cheers from the adults and over-excited yelling from the children. Somebody shouted, 'Now that's *experimental* physics, John,' to be met with the faint rejoinder from far down the garden in Oakhill's unmistakable Aberdeenshire accent, 'Aye, but it took Newton's laws to get it there!' The display was under way.

For half an hour, Lucy abandoned herself to the spectacle, the excitement, the appreciative whoops and applause, the smell of the gunpowder smoke, reliving for precious bittersweet moments those Guy-Fawkes parties of her childhood when her school friends would come round to her garden and her Mum would produce hot mince pies and her home-made lemonade and her Dad would be in charge of the fireworks and distribute sparklers with the time-worn admonition to be careful. Then, all too soon, it was over, the audience cheering and clapping with the final starburst, Oakhill, followed by his helpers, walking back to the terrace, stopping to greet newcomers and acknowledging their encouragement and thanks, finally reaching Lucy's group and speaking to them all in turn, coming at last to Lucy herself.

'John, this is Lucy,' said David, saving Oakhill the awkwardness, Lucy realised, of admitting he didn't know who she was.

'Oh, but I remember Lucy very well,' he said, looking at her directly, a broad smile crinkling his face mischievously. 'You are the physicist who asks the deep questions! I'm delighted you were able to come.'

'Thank you very much for inviting me, Professor Oakhill,' she managed.

'You're very welcome. And remember, it's John!' he added, chuckling, moving on to the next group.

Just as Lucy noticed Wilgoss re-entering the house and thought to herself that she would be happy to stay out here on the terrace, they were joined by a Mediterranean-looking

179

gentleman who proceeded to shake David warmly by the hand, although it was evident from the repartee that they saw each other daily.

'Lucy, this is Christos, a fellow theoretician, a postgrad, although he often passes for a lecturer! Christos, this is Lucy, my friend.'

'Ah, David is very fortunate to have such a friend,' said Christos, taking her hand and then, totally unexpectedly, bowing to kiss it. Never in her life had anyone kissed her hand! 'Please, are you in the University?'

'I'm a new student,' she said, still revelling in being called David's friend, still startled by the hand-kissing gesture. 'I mean, I'm a first-year undergraduate. In physics.'

Mike, one of the group to whom she had been introduced earlier, asked her, 'And do you think you've made the right choice, to do physics?'

'Oh, yes, it's been a long-time ambition. My dream job would be physics research.'

'You know,' Christos began in his ponderous way, 'there are six reasons for working. They are, not in any particular order, you understand, fun, respect, friends, recognition, career – how many have I given you?'

'Five,' said Mike.

'Yes, there is one more – an income,' finished Christos. 'But which of these motivations drives you?' he asked her.

'Isn't there another reason, too,' cut in Roger, another of the postgraduates. 'There's the wish to be remembered by society after you're dead, to show that your life meant something. In physics, you have a chance of that.'

'But very few physicists become famous, become an Einstein or a Marie Curie,' said Lucy.

'Yes, but you can still leave a legacy, an imprint,' said Mike. 'Just having your name on a physics paper gives you a kind of immortality; it shows the world what your contribution was.'

180

Lucy shot a glance at David who returned it with a barely perceptible twitch lifting the corners of his mouth. They had agreed not to tell anyone about submitting their paper, largely, Lucy assumed, to save face if it were rejected, not that she had any track record, of course, but she could understand how David might wish to keep their efforts quiet until they were successful.

'Of course, this is not immortality, not even vicarious immortality,' pronounced Christos.

'And in English, please, Christos,' said one of the group.

'By this, I mean that you may be survived by your physics paper, and if this paper is to survive for an indefinite time, then sure, you live through your paper – it is a vicarious existence. But the paper will not survive for an infinite time.'

'Sure,' said Mike, 'paper disintegrates, computer storage systems become obsolete, but if the idea is important, the idea will survive, and if it's important enough, it will be linked with your name for as long as there is civilization.'

'Yes, look at Euclid, look at Archimedes, Plato,' said Roger.

'All Greek mathematicians, of course,' said Christos, winking, looking around the group. 'But,' he continued, 'these men lived very recently. Less than three thousand years ago, they were alive. But of those who lived three *hundred* thousand years ago, we have no knowledge, and we can never have such knowledge.'

'Yes, but that's an artificial distinction,' said Mike. 'We have language now; we have ways of preserving knowledge that weren't available in pre-history. But, in principle, it's possible to reconstruct information about who or what lived as far back in history as you care to look.'

'Ah, now we come to the interesting question!' Christos was enjoying himself. 'I suggest to you that it is no more possible, even in principle, to be sure about specific details in the past than it is to go back in time.'

'Well, sure it is,' said Mike. 'You just have to calculate the reverse of the current directions of all the particles in the universe to find out what was there before.'

'I believe that would not work for two reasons,' said Christos. 'Classically, even the smallest errors in such back projection would accumulate to significant errors eventually – it is just chaos theory. And, quantum mechanically, you would never be sure exactly when and where an interaction would happen. Again, the errors would accumulate.'

'Yes, but small errors would still give you the correct picture overall,' said Mike.

'Ah, Mike, you have made my point for me,' said Christos, beaming. 'My point is that, yes, you would be able to deduce a broad picture, but the picture would lose detail as you go further back into the past. You would never, for example, be able to reconstruct the thoughts of a specific hominid that lived a million years ago.'

'That's an interesting point you're making there, Christos,' said David. 'What you're saying is that the detailed past is as uncertain as the detailed future. We can only really know about the present.'

'And we're not much good at that either,' said Roger to laughter.

'Of course, eventually, statistically, you could project back to the beginning of the Big Bang,' put in another postgraduate student. 'By then, you would have lost all detail, but you would end up, whatever meandering route you took, with the Big Bang singularity.'

Lucy's attention had begun to drift, the voices receding in her head, not because she lacked interest in the argument, but because the discussion had sparked a separate train of thought in her mind which she began to follow. Unlike the rest of the group, since their discovery of the Inverse Gödel Theorem and the God System which it pointed to, she and David had had the benefit of picturing the universe from that loftier perspective.

182

Surely, from the point of view of the God System, the mathematical system of the universe must describe the universe throughout all space and at any time? Within the God System, you wouldn't have to wait thirteen billion years to find out what the present universe looked like, because time was simply a product of our own particular universal mathematical system.

No, the God System had to operate independently of the dimensions that one of its subsystems had created! So, did that mean that the complete history of the universe, as well as the indefinite future, was already available within the God System?

This disturbed Lucy, because that might require an unimaginably huge, multi-dimensional matrix to exist within the God System for each universe generated by all of the different mathematical systems that were subsets of the God System. Or, perhaps, the basic equations of the mathematical systems could be used to reconstruct any part and epoch of their associated universes. Perhaps these two pictures were just two different ways of saying the same thing.

Either way, where did that leave free will? If a person's thoughts and actions were entirely computable, albeit in a higher-level meta-universe, or even set out already in a vast many-dimensional mathematical library of parallel universes within the meta-universe, then where was the element of choice? It would certainly cast the Old Testament God in a poor light for wreaking vengeance on His poor subjects for following an inevitable downward path to proscribed misdemeanours. She could imagine David arguing that an individual would be unable to predict their own future with any certainty because of the impossibility of omniscience of one's own universe, but she was still troubled by the implications of the concept.

She knew what David would say – he would point out that it is impossible to know the structure of the God

System or how it worked. Nevertheless, he hadn't said that you couldn't hit upon the information by chance – only that you couldn't prove you were right if you did! So that didn't prevent her from speculating!

She supposed that Christos must be right – in this universe, you couldn't reconstruct every detail of the past by extrapolating back from the present – it would be just as impossible as projecting forward in time with one hundred percent accuracy. Indeed, she could imagine David marshalling the same argument that he had used to prove there couldn't be an omniscient being in the universe. But – dare she even think of the upside to this concept? – if the whole history of the universe were already written in the meta-universe, could any part of it be rescued – *resurrected* – in the God System – even including self-aware minds?

She realised, naturally, that her wishes were racing ahead of the logic that she should be applying as a physicist. She was objective enough to acknowledge that she was increasingly identifying the God System with her idea of God, although, of course, the God System was not conscious, not personal. She would have to be careful if and when she raised these ideas with David: she wouldn't want to go down in his estimation for wild, unfounded speculation!

The group had moved on to discussing work that one of the students, Mike, had apparently been doing for Wilgoss. While she was grateful to David for looking after her so far, she wanted to show him that she could be independent, and this seemed to be an appropriate juncture to circulate a little. So she moved out of the group and headed for the kitchen, picking up a paper plate, filling it with salad, quiche, cheese and other finger foods that Mary had prepared, refilled her wine glass and walked back into the large hallway.

It was a large house, as David had told her: looking around her, she counted five doorways apart from the kitchen. Three of these were open with hubbubs of conversation emerging from each of them. She entered one at random, spotted a vacant couch and claimed one of its two seats so that she could balance the plate on her knees and put the drink down onto the table beside the couch. Almost immediately, one of the people in the room came over to her couch and sat down in the seat beside her. It was Wilgoss.

Her first instinct was to pick up her drink, rise from the couch and move out of the room, but she dismissed it immediately: if Wilgoss hadn't noticed her, he certainly would if she were to get up and leave as soon as he had sat down. No, she had to sit tight and maybe he wouldn't start up a conversation – after all, as David had said, he wasn't particularly sociable.

'Excuse me, young lady, I thought I knew all the postgraduate students, but I don't believe we have met,' he announced in the unforgettable voice that she had listened to for his hour-long lecture, causing her to relive those excruciating final moments of her public humiliation. At least he didn't seem to recollect seeing her before.

'Actually, I'm not a postgrad,' she said, her voice sounding strained and high-pitched, to her annoyance. She cleared her throat. 'I'm actually an undergraduate. Physics. I'm a physics undergraduate.' *Please don't ask me which year.*

'And which year are you in?'

Damn! 'I'm actually in the first year,' she said, irritated at not being able to think of anything to add, vexed by her own inane attempts at conversation.

'Are you, actually?' (*He was ridiculing her!* Her annoyance with herself flipped over into anger directed towards Wilgoss, and at that moment she realised she was no longer afraid of him.) 'And may I ask your name?' he persisted.

185

'My name's Lucy Darling. And you are?' She wouldn't give him the satisfaction of thinking himself so well known as to need no introduction.

'I am Professor Wilgoss. But forgive me, are you connected with young Lane by any chance?'

'Yes, I came here with David tonight.'

'Ah, then we have two things in common.' He paused, waiting for her response, but Lucy was not going to play. 'David Lane and co-authorship of the same paper,' he explained. 'I understand that the idea for the paper was yours?' This was said with a deprecating little laugh – a private joke to be shared only between themselves: she couldn't possibly have come up with the idea herself.

'Well, of course, the paper is a shared effort – we all contributed,' she said, watching his eyes, but he didn't seem to register the irony. 'My input was to suggest that the paradox presented by the undecidable proposition of a Gödel statement in any sufficiently complex mathematical system may be resolved by postulating a higher system in which the statement is provable when applied to the original system.' *Yes!* Her adrenalin had whipped up the jargon in her mind, the appropriate vocabulary bubbling to the surface of her attention so that the right words came out smoothly, confidently, without hesitation, and, as far as she could tell, in an accurate and succinct description of her contribution. Certainly, if David was right, Wilgoss, unfamiliar as he was with the field, would be unable to find fault in her deliberately grandiose sentence.

'Of course,' said Wilgoss, seemingly undaunted by her mastery of the subject, 'the idea is an obvious extension to Gödel's theorem, but you are quite right: it's time someone put it down in print. We are indeed indebted to you for that.'

You had to admire his chutzpah! 'So what areas of physics are you particularly interested in?' she asked, trying to find a neutral topic where he couldn't needle her.

186

'A very good question if I may say so,' he replied, his eyebrows raised as though surprised that she could be so intelligent as to have enquired about his work. 'I have many interests – all of them in theoretical physics, I might add. For instance, right now I am working on a famous question in quantum mechanics about the momentum of a photon in a medium. My greatest interest lies in cosmology – particularly gravitational physics.'

Lucy forked a wedge of quiche into her mouth, and smiled slightly, nodding for him to continue.

'You can do profound physics with some simple thought experiments,' he went on, 'even with something as basic as a rotating bucket of water. Indeed, I gave a symposium on that very topic last month—' He stopped, peered forward at Lucy and she suddenly felt sick. 'Wait a second – were you there?'

'Yes, I was there,' she said in a small voice.

'And was it you who asked the question at the end?'

'Yes, that was me.'

'Extraordinary coincidence!' he exclaimed, more to himself than to Lucy, looking away from her and out into the room, absorbed in the recollection.

Abruptly, he turned back to her, looking at her directly. 'Young lady – Lucy – I owe you a deep apology!'

Lucy was dumfounded. This was so unexpected that she said nothing, didn't even nod or smile, but just stared back at him, wide eyed.

Wilgoss continued: 'After the lecture I was wrought with anguish that I had put you down so, and quite unable to make any amends. I have no excuse whatsoever, but you may wish to know that the reason for my words, which were no doubt deeply wounding to you, was a certain pressure which I was momentarily unable to contain: you were simply the unfortunate recipient who happened to be in the firing line.'

Still stunned, Lucy was able to grasp only the gist of his words: she was expected to say something now. 'That's all right, Professor Wilgoss,' she said finally. 'It's kind of you to apologise, and I accept it of course, but there's really no need for you to apologise.'

She was horrified to see Wilgoss's eyes brimming brightly. 'You are being too nice to me. I don't deserve to be so lightly let off,' he said to her. Shockingly, she could feel her own eyes prickling in sympathy.

'Please,' she said, 'it doesn't matter. But thank you for explaining what happened – I can see that it wasn't personal.' She scrabbled around for something to say that would take them away from this dangerous topic, moving them onto safer ground, and then she hit upon it. 'David said you're his PhD supervisor?' she said lightly, trying to change the mood.

'Yes, I have that pleasure. Not that he needs supervising – he's very much self-propelled – but then, you probably know that already.'

'Yes, I've seen him in action!'

'Well, it would be a comfort to think that he will shortly complete writing up his thesis. A kind of closure, if you like.'

What a curious way to put it! 'You gave him his thesis topic,' she said. 'It must be hard to come up with something new for every student, especially in theoretical physics.'

'Perhaps surprisingly, it's probably a lot easier than it must have been before the nineteen-seventies,' he said. Lucy was relieved to see him become more animated now, less introspective. 'Many people thought, like Hawking famously did initially, that the end was in sight for theoretical physics – he reckoned that by the end of last century we should be doing little more than mopping up. And then there came what turned out to be an almost exponential growth in the number of new and, in some cases, completely wacky ideas.'

188

'Surely, after the nineteen-twenties, the nineteen-thirties, there must have been large numbers of problems in quantum mechanics and general relativity?' Lucy suggested.

'Ah, that's the very point. The problems were mainly confined to just these topics. But now, physicists have all sorts of ideas, many of them widely diverging from these two old stand-bys. For instance, you have the holographic principle, which your David is working on,' (her heart jumped at the phrase) 'brane cosmology that comes out of superstring and M-theory, gravity as an emergent property of entropy, dark energy, dark matter, the fifth force, loop quantum cosmology, new challenges to general relativity like Lorentz symmetry breaking – I could go on – and then you have all of the subtopics within these topics, Calabi-Yau manifolds, E8 lattices, closed time-like curves, and so on, and then, of course, we still have that good old staple, the Theory of Everything.'

For a moment, Lucy was unable to respond, caught napping by the verbal barrage that had streamed from his lips – and then she ventured, 'Of course, if you apply the principles of our paper, you'd never know for certain whether you'd hit upon the true Theory of Everything.'

Wilgoss smiled enquiringly, waiting for her explanation.

'Well, our paper's just the beginning. David's going to apply it to the mathematical system that underpins the universe. A fallout from that is that there are things about the universe that you can't prove. So, while you might be lucky and guess the correct Theory of Everything, you couldn't prove it – not all of it, at least, assuming that a Theory of Everything is really meant to explain everything in the universe.'

She felt the oddness of saying this to someone, let alone Wilgoss, when it was exactly the sort of thing that David would have said to her.

'David intends to apply it to the universe, you say?'

'Yes, and the really exciting thing is that, once you accept that the universe is purely mathematics, then it becomes clear from the paper we've just written that there must be a hierarchy of meta-universes above our own universe. Privately, we call the ultimate system the *God System*.'

Wilgoss did not match her smile. She slapped a hand to her mouth, guilt showing in her eyes as she looked Wilgoss. 'Oh Lord, David didn't want me to tell anybody about its application.'

Wilgoss said quietly, even coldly, sending a shiver through her, 'You may depend upon it that I shall share this information with absolutely nobody.'

'I'm sure he didn't mean you, Professor Wilgoss,' she said, taken aback by his sudden change of demeanour.

'Indeed, I believe you are right,' he said in the same quiet, level tone. 'Nevertheless, you can rest assured that I shall keep my own counsel on this.' He jerked his left arm out to expose his watch and examined it, holding his head back to help him focus. 'Oh, it's later than I realised. I'm afraid I must be going now. Do enjoy the rest of the evening.' With that he set his cardboard plate down on a stool beside him, still loaded with party food, rose awkwardly out of the couch and ambled out of the room before Lucy could collect herself sufficiently to return his farewell.

She looked down at her plate and began mechanically to eat some of the quiche, hoping that nobody had seen Wilgoss's abrupt departure from her side. She just wanted time to replay what had been said and to try to make sense of Wilgoss's behaviour. Most of all, she was greatly afraid that she had let David down by revealing their idea about applying the Inverse Gödel Theorem to the universe, and she didn't look forward to telling him.

One of the postgraduates came into the room and sat down beside her and she responded automatically to the

190

predictable questions, all the time thinking about what had just happened. When her plate was empty, she left on the pretext of replenishing it, met David coming out of the kitchen into the hallway, and they moved into another room to join a small audience watching one of the postgraduates singing songs that he had written himself, accompanying himself on a guitar. She was glad of the excuse not to speak, because the environment didn't seem private enough for what she had to say, and she would hate to have to make small-talk with David, thinking all the while about what she had to confess to him.

The remainder of the evening passed uneventfully, David happy, relaxed, enjoying the company of his friends and colleagues, and Lucy joining in absently, never quite at ease in what she reckoned would, under other circumstances, have been a great party. She was glad, therefore, when the raised pitch of farewells in the hallway finally signalled the departure of the majority of the guests, flocking towards the front door, she and David joining them. David turned the car expertly in the narrow road and then they were headed back towards town, luxuriously cocooned against the cold, dark night outside.

'So, what did you think of partying with a bunch of physicists?' asked David.

'Is that the collective noun? It was great,' she replied. 'Quite a few surprises!'

'Yes? How do you mean?'

'Well, I had imagined there would be discussions on a deep, philosophical level, and I wasn't disappointed, but I guess I hadn't been prepared for how playful physicists can get, too. In a nice way, I mean.'

David laughed. 'Physicists are people, too. We like our fun!'

'There was another surprise, too. I met Wilgoss.'

191

David slapped his knee in amusement. 'Hah! Roger told me he saw the two of you sitting together like comfortable old friends. What happened?'

'It all happened so quickly, I've been trying to piece it together all night.'

'Did he know who you were? I mean, did he know it was you on the paper?'

'Oh, yes, once I told him my name. He said the Inverse Gödel Theorem was an obvious result.'

'How typical of him! I hope you didn't let him get to you?'

'I suppose I did a little. And then he remembered it was me who asked the question at his lecture.'

'Oh, no, poor Lucy! I could've sworn he wouldn't recognise you. That must have been so embarrassing.'

'Tell me about it! He didn't remember me at first, but somehow we ended up discussing that very lecture and I guess that triggered it – that and speaking to me, I suppose. Anyway,' she added, 'you'll never guess – he actually *apologised* to me. I swear there were tears in his eyes!'

David digested this. 'How extraordinary! How out of character!' he said eventually.

'He also wanted you to finish your thesis. He said something about *closure*, which I thought was odd.'

'Funny enough, he was on about that to me, too, saying I had to keep at it to finish it.'

'And, David, there's something else, and you're not going to like it. I'm afraid I've let you down.'

'Lucy! Of course you haven't! Whatever do you mean?'

'I told Wilgoss about us applying the Inverse Gödel Theorem to the universe. I'm afraid I even mentioned the God System.'

David remained silent as they watched the continuous ribbon of road appear in the headlights and unravel towards them to be consumed smoothly, almost silently, under the stately bonnet.

192

'David, I'm so sorry, it just kind of slipped out.'

At last, David responded. 'What did he say?'

'I told him I hadn't meant to mention it and he gave me his assurance that he wouldn't talk to anyone about it. He was a bit funny about it, actually, as though he disapproved. He left right after.'

David sighed, not out loud, but she heard nevertheless, and she felt wretched at being the cause of his disappointment. 'Lucy, don't worry about it,' he tried reassuring her. 'Anyway, I was more concerned about not putting it in the paper than not telling Wilgoss. Please, it's really nothing; just forget about it.'

But she couldn't help thinking about it all the way back to New Hall. The music system switched to Beethoven, the haunting triplets of the first movement of the *Moonlight Sonata* matching her mood. As they crested the hill and glided down towards the lights of their town like an airliner on final approach, she considered what to say to him when he dropped her off. Before they had left for the party, she had rehearsed this bit in her mind, wondering whether or not to invite him up for coffee in her room. To ask him up would signal a new phase in their relationship. Up to that point, their seeing each other could be thought of as entirely professional, work related, with their walks and their coffees and even the personal interludes in their conversations merely as tasty accompaniments to the main course of their combined venture: cracking the ultimate problem – where did the laws come from that allowed the creation of the universe and even the whole multiverse? Even going to the party had been the aftermath of a physics lecture – it had not been David but Professor Oakhill who had invited her, after all.

Nevertheless, no matter how she tried to dress it up, those side dishes had become increasingly important to her, to which she looked forward, if she were honest with herself, even more than working on the Inverse Gödel

Theorem. She was ready to take the friendship to a new level. The importance of managing her time was undiminished, and she would be increasingly stretched towards her final year, but she felt that David was someone in whom she could confide like nobody else in her life, and, moreover, she believed he was mature and wise enough not to jeopardise her degree by being profligate with their time while she studied.

However, David had given away nothing of his feelings for her, apart, of course, from the time she had caught him standing stock-still, staring after her, but, as the passage of time blurred the memory, she was becoming less sure just how to interpret that moment. So asking him up for coffee might be a way to prompt him in an informal setting. It was two-edged of course – how would a man view being asked up for coffee into a girl's bedroom? The expectations of an experienced twenty-four-year-old man would be different from those of an eighteen-year-old virgin. On the other hand, David would behave absolutely correctly, she knew. It would only be a cup of coffee, after all!

So, in the end, before the party, she had decided to invite him up for coffee when he stopped at New Hall, provided nothing happened at the party to change her mind. That was before she had let him down. She had fallen in his expectations – he didn't need to say so – his silence had said it all, despite telling her to forget it. Now, the chasm between her inexperience and his maturity had been widened and exposed by her incompetence, and somehow, his quiet dismissal of her error, his dignity and even the high calibre of the car he drove, all seemed to emphasise this difference between them. To ask him up for coffee now would just look like a girl trying to behave as she thought an adult would, and she couldn't bear that. So, when the car drew up at the entrance to New Hall, she thanked him once again for taking her to the party and looking after her so well, quickly opened the car door and went smartly into the hall before he could see the tears forming.

CHAPTER 13

Monday 22 November – Wednesday 24 November

Nine days later, on the Monday morning, David sat alone in the office that he shared with Mike, staring at his laptop screen, his eyes unblinking, introspective, oblivious to the latest paper on the Margolus-Levitin theorem which he had just downloaded, his mind replaying the journey home from the party. He had not been with or talked to Lucy since that night, as, on the Saturday of the following week, he had once again had to visit his old home in the west of Scotland.

On the evening of the party, he had set out to collect Lucy with the firm intention of taking her back to his house (it was a house, not a flat, although he knew Lucy assumed it was a flat), where they would stop for coffee before he finally dropped her off at her hall of residence.

He had not arrived at this resolution easily. Twice, he had been on the verge of asking her out for dinner on the Saturdays when they had met, but he had foreseen a problem, which had dissuaded him. The difficulty was that, while the pretext for the meal – either an extension of their daytime project or the beginning of a personal phase in their relationship – could at first be kept deliberately vague, in the end, he would have wanted to pay for the meal, which would have signalled his desire for a new phase in their relationship. He was just not sure he wanted to risk frightening off this beautiful girl with her beautiful mind – he felt it unlikely that she would want to become closely involved with someone older than she by a third of her age.

The party, however, had offered the perfect context. If he could put it to her that they would call in at his place for coffee on their way back to New Hall, it would sound perfectly natural, unforced, and yet allow them to be alone together without the distractions of the party. He was still slightly nervous about asking her, and decided to wait until they were within sight of the town on the way back, which would make it seem unrehearsed and unremarkable.

In the event, it had all gone wrong. How he wished he had reacted better when Lucy had said she had told Wilgoss about extending their paper to the universe itself. Although he had tried to reassure Lucy that no harm had been done, his immediate concern was that Wilgoss was now in a position to develop their ideas himself. After all, Wilgoss now had the manuscript, and so he had the advantage that nobody else had, except for himself and Lucy, of being able to start work on it right away to apply the Inverse Gödel Theorem to the universe. He had been silent while thinking this through, and Lucy had obviously misinterpreted this as displeasure with her. The problem was, if he explained and told her of his worry about Wilgoss, that would make her feel even worse about her indiscretion: she might even think he was punishing her by letting her see the possible consequences of her mistake.

As they had entered the town, David had concluded that she might very well decline his invitation to his house, and ask to be taken directly to the hall instead. He decided, therefore, rather than to risk it, simply to go straight back to New Hall – she would never know of his original plan. Now, in hindsight, in the quiet of his office, he could see that he should have made light of her disclosure to Wilgoss, mildly teasing her, even, and then she would probably have been happy to go back to his house. However, he hadn't been quick enough to think of that, probably because of his preoccupation with Wilgoss.

He was startled by a knock at the door which immediately opened a little way, Wilgoss's head emerging into the room from behind it. 'Ah, David, I'm glad I caught you,' he said, completing his entry.

'Hi, Jeremy, have a seat,' said David, indicating Mike's chair.

'Thank you, I won't if you don't mind. This will only take a minute.'

'How did you enjoy the party?' David asked, not having seen Wilgoss for over a week, despite his office being just down the corridor.

'Very satisfactory as always. I am sorry to tell you that the paper has been rejected.'

For a moment David struggled with the change of topic. Surely Wilgoss couldn't mean the paper that they had only just submitted? 'I don't understand,' he said. 'It's only been three weeks since I gave it to you – that can't have been enough time, surely?'

'Ordinarily, no, you're right, but I asked two editorial board members to be referees, and they responded more or less immediately. They have turned it down.'

'What – they rejected it outright? Haven't they given us some indication of what we need to do to amend the paper?'

'No, there is no encouragement to amend the paper, I'm sorry to tell you. I find the best thing in these cases is just to move on – get your teeth into something quite different, finishing your thesis, that sort of thing.' With that, Wilgoss started to turn towards the door.

'Hang on, Jeremy. What reasons did they give for turning it down?'

Wilgoss looked at the floor. 'They say the central conclusion of the paper is flawed. They say that sufficiently complex mathematical systems cannot always point to higher-level systems because it wouldn't work if you applied that reasoning to the universe. Essentially, they are saying that you can't prove there is a higher system above the universe.'

'But the paper doesn't even *mention* applying the results to the universe,' David protested.

Wilgoss was still inspecting the office carpet. 'Nevertheless, they have ruled it out. Presumably extending it to the universe was obvious to them.' Wilgoss looked up at last. 'David, such a report from the referees rules out any hope. The editor-in-chief would claim that if two referees came to such a conclusion, then so would most of the readers of the journal.'

David was stunned, lost for words. Then he thought of something. 'Would you mind giving me a copy of the referees' reports, then, please?'

'They haven't said any more than I have just told you. It would be a mistake to read the reports – like reading bad reviews of one's performance on stage – quite discouraging. Just accept what I have said. I'm very sorry, David.' And with that, he turned and left the office leaving David staring at the door.

His first thought was to contact Lucy as soon as he could.

*

In the event, he waited for a couple of days until Wednesday afternoon. He had considered phoning her – they had each other's number in case one or other of them should be delayed in getting to their Saturday meetings – but then he realised that just phoning to tell her the paper had been rejected would not be fair to her, as she could do nothing about it. Better to wait until they could meet face-to-face and talk it through.

He knew that she had no lectures on Wednesday afternoons and so at ten to one he positioned himself outside the lecture theatre and waited for her astronomy and astrophysics lecture to finish. Shortly, a deep and increasingly loud rumble like the precursor to a volcanic eruption announced the end of the lecture and the imminent disgorgement of the audience. Then the doors burst open, discharging the students in a noisy pyroclastic flow, eager for lunch.

Lucy saw him waiting for her and came over to him, smiling and excited, making David feel guilty at being the one about to burst her balloon.

'Hi, David,' she said, pleased but clearly puzzled at seeing him standing there, 'this is a surprise!'

'Yes, sorry, Lucy. Look, are you on your way to lunch?'

Her smile faded a little: she had detected the serious timbre in David's voice. 'Yes, sure, why?'

'I have some news about our paper. Look, is it possible for me to join you?'

'Yes, that would be OK. We can take visitors. You'll have to pay, of course.'

As they walked across to New Hall, David told her about Wilgoss coming to his office and delivering the disappointing news. When they reached the hall, he waited in the foyer while she went up to drop her things in her room. Ten minutes later, they were seated opposite each other at the end of one of the long tables in the dining room. They had found a quiet spot quite easily, as the bulk of students tended to eat earlier.

'What I can't get my head around,' she was saying, piling her fork high with rice and curried chicken, 'is that both referees said exactly the same thing about extending our paper to the universe. It's not so much that they then say it wouldn't work, it's more that they even thought of extending it to the universe in the first place, when we didn't suggest anything of the sort in the paper – and both of them have come up with the same idea!'

David had been equally struck by the same improbable coincidence, but he wanted Lucy to work it out for herself, to see if she came to the same conclusion that he had done.

Lucy's fork froze half-way to her mouth, and she looked up at David, wide-eyed. 'You don't think my telling Wilgoss about extending the paper to the universe has anything to do with this, do you?'

'Well, I *have* been wondering that myself,' David admitted. 'But if there *is* a connection, then I think there's only one way that the referees could have got to know about applying the theorem to the universe.'

'What – you mean Wilgoss told them?'

'No, I don't think so, not directly. There would be no point in telling them if he didn't think it was a good idea to apply it to the universe. No, I think he must have included the idea as an addendum to the paper because he liked the idea and wanted to lay claim to it in print. The referees didn't think of applying the theorem to the universe – it was spelt out for them!'

'Would there be time to add it to the paper, send it off to the journal and for the referees to report back, all in a week?'

'Apparently, yes. I asked Wilgoss the same question. Of course, at that time, I thought he'd sent off the paper when I gave it to him three weeks ago, but he could just as well have sent it off last Monday, say, after he'd picked up the idea. He did say that the referees were Board members and had reported back almost immediately. So, yes, I think he'd have had time to add the bit about the universe to our paper.'

'If he thought it was such a good idea to claim ownership of the idea in the paper, do you think he might have taken our names off it, leaving his as the one author to get the glory?'

'Oh, I hadn't thought of that.' He grinned. 'I would never have guessed you had such a criminal mind!'

201

Lucy was concerned. 'Have you thought about challenging him?'

'Yes, but there's no proof. If he lied about it on Monday, there's no reason for him not to tell the same lie today.'

'I know how we can get proof! We can ask Paul what's-'is-name, the reporter on *Stat*, to help us.'

'Paul Evans? Help us how?'

'He's a computer hacker. He's hacked into university emails. He told me when he saw me for my side of the story – you know, when somebody saved my life!'

David couldn't help grinning again. 'Yes, I know about Paul's hacking predilections – he's famous. You know, you really are criminally minded. You're suggesting that we hack into Wilgoss's emails?'

'If Paul found that Wilgoss had changed the paper and submitted it under his own name, well, that would be worse than us just opening his emails.'

David felt doubtful. 'I'm not sure about this,' he said.

'Well, why don't we at least ask Paul for advice – see what he says?'

They were silent for a moment while they cleaned off their plates.

Eventually, David capitulated and they decided to see Paul that afternoon if he was available.

They walked across the campus to the Computer Science Building where Lucy had been interviewed by Paul, and asked for him at reception.

'Well, well – you two inseparable now?' asked Paul, as he came up to them, shaking their hands in turn. 'You want me to do a follow-up story – you know – how the near-death experience changed both your lives forever?'

'Not exactly,' said David. They were following Paul now, walking through a network of corridors. 'We have a problem that needs your expertise.'

'Now I am intrigued! Come in, sit down. Oh, you can't. Wait a minute.' He popped out of the office and returned with a chair which he placed just inside the door for Lucy and then he and David tiptoed across the mess of books, papers and cables to the other two chairs. 'So, how can I help?'

'Lucy tells me that you're good at hacking into computer systems, emails and that sort of thing,' David began.

'Are you asking me to hack into somebody's emails? That's pretty heavy. You must have a very good reason even to risk asking me. Who is it?'

'I'm not actually asking you to hack in,' said David. 'We'd just like your advice. It's a lecturer in physics. A professor, to be accurate.'

'And what would be the reason for hacking?'

David took a deep breath. Paul was a reporter, and so he was taking a chance in telling him. On the other hand, Paul would have to be convinced that the risk was worth taking, and, in any case, a reporter never revealed his sources if they wanted to keep a low profile. David would just have to hope that Paul observed these journalistic principles.

'We think that this lecturer has altered a paper that Lucy and I gave him and submitted it to a journal under his own name and not ours.'

'What journal is that?' Paul asked.

'The *International Journal of Natural Philosophy*,' said David.

'It must be a pretty important paper for you to go out on a limb like this.'

'I don't think I'd be exaggerating much,' said David, looking over to Lucy for support, 'if I told you that this paper could be the key to the universe.'

'I know that sounds melodramatic,' said Lucy, 'but, oddly, what David just said is, in fact, true. It proves that there exist higher systems than the universe itself.'

'OK, so who is this lecturer, this professor?' asked Paul.

'Jeremy Wilgoss,' said David. 'Will you help us?'

'Yes, I'll help you,' said Paul, 'but I'm afraid not in the way you're hoping.'

David looked blank.

'I'm going to give you a piece of advice, which, if you follow it, could save you from being sent down. Don't even *think* about hacking into anybody's email. I certainly won't do it for you. Yes, I'm guilty of hacking in the past, and I'm lucky not to have been thrown out of the university, or maybe even into jail. David, take my advice, and find some other way.'

David felt distinctly uncomfortable and rather shamefaced at receiving such an admonition in front of Lucy. Nevertheless, he could see now that Paul was giving him good advice, and was suddenly thankful that Paul had brought him to his senses, had given him some perspective.

'You're absolutely right, Paul, thank you for your wisdom,' he said, gratefully. 'I think I was just too close to the problem. Your cold shower is just what I needed!'

'Not at all, David. I'm just glad if I've prevented you from making the same mistake I did.' Paul rose from his chair, signalling the end of the meeting, and David and Lucy followed suit.

'Let me guide you back to the reception,' said Paul, and he led the way to the entrance where they shook hands and David and Lucy set off for his office.

'I'm sorry about that,' said Lucy as they walked over to the Physics Building.

'That wasn't your fault.'

'No, I mean for suggesting hacking into Wilgoss's computer. I can see Paul was right, now; it would have been a crazy thing to do.'

'Oh, don't worry. We were both caught up in the idea and we hadn't stopped to give it enough thought.'

'Is there anything else we can do?' asked Lucy.

'About the paper? Maybe there is. I could try contacting the journal to see if the paper was really in all our names and to see if Wilgoss did add something to it. Or maybe I could get a copy of the referees' reports.'

'Why couldn't you do that before?'

'Well, strictly, they shouldn't discuss the paper with me. I'm not the corresponding author, and all the contact is done via Wilgoss's computer on their website. They'll have given him a password. But now that we've ruled out hacking into his computer, I guess there's nothing to lose by trying the journal just in case.'

They entered the Physics Building and David led the way upstairs to his office. Mike was out, presumably down in the laboratory, and Lucy sat in his chair while David switched on his laptop and started to prepare some coffee.

'So this is where you do your deep thinking,' said Lucy. 'It's a lot tidier than Paul's room.'

'Yes, well I don't have the computer paraphernalia that Paul seems to have, and Mike, thank goodness, keeps his stuff down in the laboratory. We really just use the room for getting on the internet and writing our theses.'

'Where do you store your thesis? On your laptop?'

'Mostly in my head at the moment: I've written very little so far. But yes, I keep a copy of what I have written on the laptop, a copy on the university's computer and I also keep a fairly up-to-date copy at home on a memory stick.'

The laptop had fired up now and David located the website of the *International Journal of Natural Philosophy*, whose headquarters were in London. He swung the screen round so that Lucy could watch.

'See what I mean? People submitting manuscripts have to register and they get a password. We don't know what it is, of course, so we'll have to call them on the phone.'

The water had boiled and David stopped to pour it into two cups. 'Sorry it's instant,' he said, 'but anything else would delay the speed of the caffeine fix.'

He swung back to the computer, found a contact phone number for manuscript submissions and dialled it on his office phone.

'Hi, my name's David Lane,' he said when they answered. 'I'm calling about a manuscript we submitted recently.'

'Are you the corresponding author?' asked the man.

'No, I'm afraid he's not here right now, and it's kind of urgent.'

'Look, I'm sorry, but all queries about manuscripts in progress have to be submitted through the website by the corresponding author.'

David looked across at Lucy and made a gloomy face. 'Well, it's not in progress, actually: it was rejected. I just need a copy of the referees' reports' – crossing his fingers for Lucy's benefit – 'so I can work on revising the paper.'

'The trouble is, Dr Lane, that I would need to establish that you're the rightful author of the paper before I could send you the referees' reports – you see the problem?'

'Oh, absolutely. Look, if I email you the manuscript just now, surely that would show that I'm a co-author. Nobody else would have access to the manuscript apart from the authors.'

There was a pause, then, doubtfully, 'Well, could you tell me the authors' names so I could check it?'

'Yes, the authors, in order, are David Lane, Lucy Darling and Jeremy Wilgoss.'

'OK, just a moment, please.'

David took a swig of coffee. It was nice to have been called *doctor*.

'Right, Dr Lane, here it is. Ah. Did you say your paper was rejected?'

'Yes, I'm afraid so.'

'Well, not according to this. Your paper was accepted with full recommendation to publish by both referees but then it was withdrawn.'

'Eh? Sorry, I don't understand you.'

'The paper was withdrawn by Jeremy Wilgoss. The paper was received by us on the first of November—'

'Hang on, just a minute,' said David, rummaging for a piece of paper, writing down the dates. 'OK, sorry, go on.'

'—the referees accepted it outright on the tenth of November, but then we were surprised when Jeremy Wilgoss withdrew it without explanation on the twenty-second.'

'OK, I see now, there's been an awful mix-up,' said David, trying to sound as though he understood everything. 'Look, I'm just firing off an email to you right now with the manuscript. If you agree it's the correct one, could you possibly send me back the two referees' reports, please?'

Doubtfully, again: 'Well, OK. Don't use the journal's email address, use mine instead,' and David typed it into the address line and sent it off.

'You should have it now,' said David.

'No, not yet. Oh, here it is.' There was a pause; David looked across at Lucy and smiled, and she put on a puzzled frown, opened her hands, palms upward, shaking her head in bafflement, and returned his smile.

'Yes, are you there, Dr Lane?'

'Yes, still here.'

'Well, yes, your manuscript is identical with the one that was submitted, and so I can see no reason not to send you the referees' reports. I'm sending them now.'

'OK, got them,' said David a moment later. 'Look, thank you very much indeed for your patience. We'll get ourselves sorted out at this end very shortly.'

He put down the phone. 'What on earth was all that about?' Lucy asked.

'I'm still trying to get my head round it myself. It seems that Wilgoss sent off the manuscript on the day I gave it to him, and then,' he looked down at the dates, 'both referees accepted it just over a week later.'

'They *accepted* it?'

'Yes, and I'll just print off their reports now.' He opened the files that he had just received and pressed the print button.

'Then,' he continued, 'on the twenty-second, nearly two weeks after the referees accepted it, Wilgoss withdrew the paper.'

'What did the referees say?'

David fished the sheets from the printer's out-tray, gave one to Lucy and started to read the other one.

'Mine's excellent,' said David, finally, swapping his report with Lucy's.

'Yes, so's mine.'

'Well, I guess there's only one thing to do now,' said David when he had read both reports.

'See Wilgoss? Should I come with you?'

'Probably best not,' said David. It mightn't be good for Lucy if Wilgoss knew she had seen him being caught in a lie, which was almost certainly going to happen. 'I think I should see him alone.'

'Will you let me know what happens?'

'Lucy, of course! I'll call you later.'

So Lucy set off back to New Hall, leaving David to psyche himself up for the confrontation to come.

*

David hesitated outside Wilgoss's door. Wilgoss was rarely to be seen about the building; he seemed almost to live in his office, sitting at his desk apparently staring into space for hours at a time. This was well known amongst the postgraduate students, since the office window of one of them faced across the quadrangle directly into Wilgoss's room, where he could be seen, motionless at his desk, illuminated by his office light far into the night.

David had tried to rehearse how the conversation would go, but had abandoned the idea: it was like a chess game with too many possibilities burgeoning just one or two moves into the game, and so, before he could reconsider, he knocked and waited for the faint 'Come!' from within the room.

'Do you have a moment, Jeremy?' said David, peering half-way round the door, trying despite himself not to imitate Wilgoss's style of entry.

'Yes, David, what is it? Please sit down,' he added as an afterthought, gesturing to one of several chairs he used for tutorials with the senior undergraduates.

David decided it would be less painful to get straight to the heart of the matter. 'I phoned the *International Journal of Natural Philosophy* to ask them for a copy of the referees' reports of our paper.'

Wilgoss's face was impassive. 'And they wouldn't give them to you.' It was said as a statement, a foregone conclusion.

David looked down and then back up at Wilgoss. 'No, actually, they did,' he said, his voice quiet, level, in the same tone that he would have used to agree the nights were drawing in.

Wilgoss held the stare for five seconds and then it was his turn to look down. 'Ah,' was all he said.

David waited, still looking at Wilgoss, but not holding his head high, not challenging him. In the final analysis, regardless of who held the moral advantage, it was an unequal meeting: David was a student, albeit a postgraduate, and Wilgoss was a professor of theoretical physics.

Eventually Wilgoss looked back at David, his eyes tired, even sad, the wrinkles round his forehead and face reminding David for the first time of a bloodhound's saggy demeanour. 'You want to know why I withdrew a perfectly good paper – *significant*, I think was the word the referees used – and why I … misled you.'

'I just want to understand, Jeremy.'

Wilgoss carefully straightened his right leg, rotated it slowly round and bent it again in his characteristic manner. 'When the referees reported back, I was very pleased for you. As luck would have it, or perhaps God, who knows, I didn't see you to tell you the news before John Oakhill's party some days later. That was when Lucy Darling told me that you were applying the work to the universe itself.'

Wilgoss twisted himself slowly round in his chair, sticking his other leg out now and then folding it back under his chair. 'I did not do this lightly. I thought about it over the weekend, and I could see no flaw in the reasoning. Lucy was right – if you apply this to the universe, it goes all the way. I agonised for a week before I withdrew the paper on Monday. I came and told you right away. That the paper was not to be published, I mean.'

'I'm sorry, Jeremy, I don't understand. If you agree that the paper points to a higher-level system above the universe, then why not let it be published?'

'Because of the way it would be misinterpreted. If the popular press got hold of it – and they would – it would be seen as a scientific proof of God. Lucy told me you even refer to the higher-level system as the *God System*, facetiously, no doubt. But, if two scientists are doing that, imagine what the general public will make of it.'

'You're right, Jeremy, we're not serious when we call it the *God System*. After all, I'm an atheist! Anyway, the higher-level system is hardly the traditional image of a god – it's not even conscious for one thing.'

Wilgoss smiled without warmth. 'Oh you may be surprised what's in your paper!' He stretched out his right arm, slowly twisting it as though to get the blood flowing in it. 'Unlike you, I am a Christian, and this is not the way to God. Jesus said "I am the way, the truth, and the life: no man cometh unto the Father but by me." That rather rules out manipulating Gödel's Theorem to get there.'

For a moment, David thought Wilgoss was joking, and then, with something of a shock, realised that he was absolutely serious. 'Where does this leave us, then?' he asked.

'Oh the cat's out of the box, now,' said Wilgoss, and he sighed deeply. 'This will run to its inevitable conclusion. I suppose the least I can do for you and Lucy will be to contact *Int J Nat Phil* and ask them to ignore my withdrawal of the paper. I'll take my name off it, of course. I'll copy you in on the email.'

'Thank you, Jeremy.'

'David, I hope you realise that none of this was personal. You're a good physicist, and I wish you every success.' With that, he rose, and David, realising this was the end of their meeting, got up, too, and went to the door.

'Thank you again, Jeremy.'

'Good-bye, David.'

David closed the door quietly behind him, walked back to his office and sat down, his mind buzzing with conflicting emotions. He was not entirely surprised to find his eyes prickling with tears, although with relief, excitement, disillusionment or sadness, he could not say.

CHAPTER 14

Monday 29 November

The following Monday morning, when David arrived at his office at ten o'clock, he was still on a high. Wilgoss, true to his word, had copied him an email to the journal soon after he had returned to his room, and David had promptly followed it up by calling back the same man he had spoken to in the editorial office half an hour before. The office had agreed to go ahead with publishing the paper under his and Lucy's name only, just as Wilgoss had instructed in his email to them.

David had been almost euphoric when he phoned to tell Lucy later that same evening, and they had arranged to meet three days later, on the Saturday afternoon, by the eighteenth hole of the golf course. From there, they had wandered for over a mile along the wide, pale-pink flats of the deserted sandy beach bordering the golf course, both of them insulated by their thick anoraks against a biting easterly wind blowing in from over the grey sea, both engrossed in working out the approach to their next paper, spurred on by the spectre of competition from other physicists in a race that would surely be triggered by publication of their paper, the starting gun.

He was still reflecting on the progress they had made on that walk and in their subsequent discussions upon returning to the enveloping steamy warmth of the Coffee House when Mike came into the office, looking as though he had just emerged from sleep.

'Good weekend, then, I see,' said David.

'It'd be fine if I didn't have to come in so early for my experiment.'

'Coffee?'

'No, I need to go down and switch on the plasma cell. Maybe after.'

'You go down,' said David, 'and I'll follow you with your coffee. I'll put it on now.'

'Cheers, mate,' said Mike, picking up a folder and leaving for the laboratory.

Three minutes later, David had managed to stack Mike's mug on top of his own, both full of hot coffee, leaving one hand free to open, shut and lock their office door. Then, holding the mugs with both hands, he made his way down the two flights of stairs to the bottom corridor that ran all the way round the building, giving access to all of the laboratories. He balanced the mugs in one hand again, opened the door to Mike's laboratory very carefully, partly to avoid spilling the coffee, and partly through long experience that you didn't go charging into any laboratory, especially not one with laser radiation at large.

When he saw Mike lying prone on the floor in the opposite corner of the laboratory, he dropped the mugs anyway, both of them shattering on the hard, shiny green tiles, the brown coffee looking like a child's drawing of a bomb-burst, its thin fingers radiating from the shrapnel of broken pottery spread in a halo around the epicentre.

'Mike, are you all right?' he called out, running over and crouching down beside him, feeling momentarily light-headed himself. *Was Mike breathing?*

He was lying on his front, right arm extended, left arm by his side. His head was on the floor, turned to the left, and just as David was putting the back of his hand below Mike's nostrils, he stirred, drew up his knees, and tried to get up.

'No, stay where you are, Mike,' David told him, placing a hand on his shoulder. 'Just lie still for a moment.'

213

'Gas,' said Mike, mumbling, so that David could hardly make out the word.

Still crouching, David glanced to his side, where the door of the gas-cylinder store stood ajar. He registered, without understanding for the moment, the shards of black sticky insulating tape hanging from the door jamb, more sticking up from the sill, and then he saw them, weakly illuminated from behind the door by the low-wattage light of the store-room, on the floor, two feet clad in black socks, shod in black shoes, and the cuffs of dark-grey striped trousers protruding from behind the door.

Still feeling slightly giddy, David rose and pushed on the door. It gave, only slightly, but enough to let him put his head through and see, sitting up, wedged in the corner behind the door, his head lolling back and his mouth open, his feet straight out in front of him, preventing the door from opening further like a grotesque sausage draught excluder, the body of Wilgoss.

'Oh Jesus!' he cried, springing with a reflex jerk back into the laboratory. He had little experience of dead bodies, but Wilgoss was clearly dead: his eyes had been fish-like, clouded over, staring from the corner of his terminal retreat.

Mike was propped up on one elbow now, and David sat down beside him, not trusting himself to keep standing upright.

'What happened, Mike?' he managed to ask.

'I think we've just been hit by a big bubble of gas. I just went to turn on the supply for the plasma cell. I had to push the door really hard.'

'Yes, Wilgoss's body was behind it.'

'Not just that,' said Mike, visibly recovering by the second. 'The door had been sealed with that tape. From the inside.'

'So the gas is gone, now?'

'Yeah, probably dispersed. I'd better call Tony,' he said, rising to his feet and walking across to the laboratory phone.

214

In less than two minutes, Professor Tony Packard, the head of Mike's group, came into the laboratory. Stepping around the debris of the coffee mugs he asked the secretary who had accompanied him to remain at the door and not to let anyone in.

He went over to the gas store room, peered round the door without opening it further, and emerged, plainly shaken. 'Yes, he's obviously dead.' He pulled out his phone, could find no signal, went over to the laboratory phone, punched in a number and said 'Police, please.'

'We have to remain outside in the corridor and wait for them to come,' he said, when he came off the phone. As David and Mike made their way in silence to the exit, he called over to the secretary for the number of the Vice-Chancellor's office and picked up the phone again.

They waited in the corridor with the secretary until Packard came out and asked Mike to lock the laboratory door. 'Would you mind finding John Oakhill, if you can, and let him know what's going on?' he asked the secretary, who took off down the corridor, the efficient tap-tapping of her heels highlighting the contrast between her adrenalin level and the numbness that united the three of them left standing like lost souls outside the laboratory door. It occurred to David that she had been the only one not to have seen Wilgoss lying there.

Quite soon, they had become the nucleus of a buzzing knot of physicists and technicians as the news pervaded the building like a bad odour attracting flies. Then, just as quickly, the crowd dispersed in advance of three policemen clumping down the corridor, the protective body armour under their fluorescent yellow jackets lending them a larger-than-life presence, the three physicists just accepting the unfamiliar sight as unremarkable in contrast with what they had just witnessed on the other side of the door.

The senior officer addressed Packard: 'Are you the person who reported the death?'

Packard confirmed that he was and the officer took out a pad, noted his name, address and occupation, asked who had discovered the body, recorded David's and Mike's names and addresses, too, and asked for the key to the laboratory.

'Is there anything dangerous in there?' he asked before entering.

Mike told him that he thought there had been non-poisonous gas in the room that had displaced the oxygen, but that it seemed to have dispersed.

After a minute, the officer came out again, locking the door behind him, holding onto the key. 'Yes, unofficially, he's dead, all right,' he announced, and asked Packard if he knew the dead man. Packard confirmed Wilgoss's identity.

As one of the other officers began to cordon off the area around the laboratory door with tape displaying diagonal blue flashes and the injunction not to cross the police line, the senior officer and the third policeman herded the three back along the corridor, where they met John Oakhill hurrying towards them. When it was established that Oakhill, although not a witness, was Wilgoss's close colleague, the officer wrote down his details and asked him to remain in the building until the investigating team arrived.

On their way upstairs, they encountered more police coming down. Then they were escorted individually, each with one of the policemen, to separate rooms. David was accompanied back to his own office by one of the new policemen, no older than himself, whom they had met on the stairs; Mike had been taken elsewhere.

Although David didn't in the least feel like socialising, he felt that he needed to anchor the day in normality once more, and so he asked the policeman if he would like some coffee, realising immediately he said it that there were no more mugs in their office.

'Would it be OK for me to go out and get a couple of mugs?'

'No problem. I'll come with you, sir,' said the policeman, and they went along the corridor to the kitchen at the end, passing another policeman taping Wilgoss's office out of bounds.

As they sat down once again in David's office, drinking the coffee he had made, he began to understand the enormity of what was happening. While it had been obvious to Mike and himself that Wilgoss had committed suicide, the police treated all apparent suicides as suspicious, at least in the first instance, as a matter of course. The policeman opposite had explained to David that they had to wait for the senior investigating officer to arrive. The officer would not be able to start interviewing until he had done a walk-through at the scene of the death.

While David was confident that they would shortly be convinced that this was a suicide, what weighed on his mind was his own part in Wilgoss's death. It had only been last Wednesday, five days ago, that he had exposed Wilgoss's lie. His mind was so churned up that he found it difficult to recollect their conversation exactly. He did remember thinking it odd, at the time, that Wilgoss had bidden him *good bye*. Did Wilgoss know then that it was the last time he would see David? If so, had it been that very meeting that had made up his mind? The question that he kept coming back to, and which would now never be resolved, was: if David had not confronted Wilgoss that day, would Wilgoss still be alive?

Then a thought flashed across his mind that left him feeling sick – if the police knew about the confrontation, could they construe that as deliberately pressurising Wilgoss to commit suicide? Fortunately the policeman opposite seemed content to gaze out of the window, not inclined to chat – at that moment, David would have been quite unable to engage in normal conversation.

217

Nobody knew about their meeting except Lucy, and the police would have no reason to speak to her. Then he remembered that they had taped off Wilgoss's room. Would there be anything there? His computer, of course! The police could get the computer systems administrator to open Wilgoss's files, and they would easily find the email sent so recently about David to the journal. He clasped his hands to his head and leaned slowly forward, supporting his elbows on the desk. The policeman said nothing.

Chapter 15

Sunday 28 November

Jeremy Wilgoss sat motionless in his darkened office, watching the western sky turn from yellow to red to brick-red. These would be the last photons he would ever see coming from the Sun, he thought. At a quarter past four, it was already quite dusky outside, but his eyes had become slowly adapted to the failing light, and he could see well enough in the office. Leaving the light switched off would give the impression that his office was empty. He would rather not talk to anyone in his last hours alive on Earth, although he rarely had unannounced visitors, and, anyway, Sunday afternoon and evening was the graveyard shift in the Physics Building. What an apt expression!

He was one of very few people privileged to know the appointed hour of their death, he reflected. Privileged? He supposed that prisoners certain of the moment of their execution a few hours in the future may have felt desperate rather than privileged, but perhaps that desperation had been borne of hope – hope of a last-minute reprieve from the Home Secretary or the State Governor that might never come. In his case, there was no hope of reprieve. Maybe a closer comparison would be a terminally ill patient who had an understanding with his doctor, the doctor explaining that the dose of diamorphine he would administer to control the pain would also kill him.

He was calm now, totally accepting the inevitable. He was even excited, anticipating what would happen next, when his soul passed on. He had slight qualms that this was not how God had wanted it – this was not his time – but he had thrown his soul open to God and He would see that this

219

poor sheep lacked the bravery to stay alive. God was not vindictive; He would not punish cowardice. Even the manner of his death had been planned for the minimum of discomfort: indeed, he believed his passing would be perfectly peaceful.

He had first heard the word *Huntington's* when he was 17. Even now, after 35 years, the very name seemed to drip with dread, the unspeakable that was searching – *hunting* – for *him*! It had been four in the afternoon and he had just come home from school. He was the youngest of three brothers, the other two of whom were then both at university. At first he thought the house was deserted, but, as he was climbing the stairs to his room, he caught an extraordinary sound – his father singing upstairs, not a song, just notes, muffled as though his bedroom door was shut. His father never sang, and, moreover, his father was never home at this time – he was not due for another two hours. He walked up the rest of the stairs more slowly, the better to hear, and as he reached the landing, he froze, motionless, shocked beyond anything he could remember, as it registered in his disbelieving mind that his father was not singing but sobbing.

Now he could pick out, in chilling counterpoint, his mother's low-pitched voice, also muffled behind the door, soft, monosyllabic, soothing. He was frightened, terrified, by what must be going on behind that door. His father *never* cried. A couple of times in his life, he had heard his mother cry, but, while those occasions had been alarming, they had nothing of the power of this. He knew, instinctively, that this must signify disaster, not just for his father, but, somehow, for the whole family.

He turned and started quietly down the stairs, but the creaking of the tread on the second step gave him away, and the bedroom door opened, framing his mother. He trotted down to the bottom and into the living room and his mother followed him in and enfolded him in her arms. It was then that she told him of his father's appalling disease, how he had been diagnosed by the neurologist at the hospital with Huntington's disease. She called it Huntington's *chorea*: that was how the hospital had referred to it then. Chorea meant a dance, she said, a hideous corruption of a word with such happy connotations. Victims' brains would be progressively consumed by the disease, manifest by a slow, writhing, tortuous movement of the limbs and head in a terrible *danse macabre* over which the victim had no control.

She was not finished. She carried on relentlessly, as though to void herself of the vile information in a single disgorgement, her words hammering into his unwilling mind, barely able to grasp the enormity of her message. The progress of Huntington's was inexorable. Sufferers gradually lost their mental faculties; dementia was the inevitable result. With the unfeeling selfishness of youth, he had asked his mother if he would get it too, knowing how this question always elicited a firm, confident, reassuring denial. To his horror, she had not denied the possibility. The disease was inherited – she would hold nothing back, he needed to know. There was a fifty-fifty chance that any child of a sufferer would also get the disease. A fifty-fifty chance for each of his brothers – and for himself.

She had tried to temper the news by telling him that, if the worst came to the worst, he would not suffer until he was his father's age, and this had helped, a little. After all, his father was an old man from his perspective. Some years before, ironically, he had imitated his father's gait at the very earliest manifestation of the disease, the early symptoms only recognisable years later in retrospect. The mimicry was not malicious, and largely unconscious – his father's walk, his legs apparently thrown forward with each step, seemed to him arrogant and powerful and he worshipped his father.

221

However, as the disease ran its unerring course, his father's involuntary movements became worse, not completely controlled even with the best medication, his mind lost its sharpness, and he retired from work. Within a year, he had been admitted to a home that catered for Alzheimer's patients: this was the first case of Huntington's they had seen, but they took him in anyway. At first, he would visit every evening, always with his mother, because his father now said very little, and the visits tended to be dialogues between his mother and himself, trying unsuccessfully to include his father, but, as the days became weeks and the weeks months, and his father stopped acknowledging them at all, staring at the moving pictures on his television, the smell of urine sharp in the room, he visited less frequently, and then not at all. His father choked to death one evening not long after that, totally unaware that he had ever been part of a family, or, indeed, what a family was.

His genetic demon was a constant presence, never retiring entirely into the background. Often, he would dream that he had actually exhibited symptoms of the disease, that he had, after all, inherited the diabolical gene, and then the momentary euphoria when he realised on awakening that this was just a dream was cruelly extinguished by the sudden brutal recognition that his nightmare stood a heads-or-tails chance of becoming reality. There was no escape. When first his eldest brother and then the middle one developed the disease, he was devastated, but, physicist as he was, who knew better, his inner voice would try to persuade him that the chance that all three brothers had inherited the disease was small. He had considered genetic testing as he got older and it became widely available, but, to him, living from day-to-day in agonising uncertainty was to be preferred to the fifty-fifty chance of being handed such a cruel death sentence.

In the end, he hadn't needed a genetic test. In the last year, he had become convinced that his walk, always a little clumsy, had begun to develop the subtle, sinister signs to which he, of all people, was exquisitely tuned to recognise. With a deep sigh, he had mentally reached down into his most secret, bottom drawer and extracted the plan that he had been carrying in his head since the day his father died. He had resolved then that he would not die a death like that.

So, over this last year, he had proceeded methodically and unemotionally to research into how he should die. Ideally, he would have chosen pentobarbital, repeatedly described by witnesses as inducing a peaceful death. However, it seemed to be available only in Mexico, and he could not be confident that they would sell him the correct strength of drug, or even the correct drug at all. In addition, it seemed that it had to be taken with another drug to prevent vomiting it up, and there was a danger of losing consciousness before you could take a lethal amount.

By chance, just as he was ruling out drugs, he happened to read a report in the town's newspaper of the death of a local welder who had been working inside a metal tank. The welder had had to use an inert gas, argon, to cover the weld and prevent it from reacting with air until it had cooled. The dense argon had collected in the bottom of the tank, displacing the air. When the level of argon had risen above the welder's face, his brain simply ran out of oxygen. Since argon was undetectable, the welder had had no warning of his lack of oxygen and had died from asphyxiation.

It was at that moment that Wilgoss had made the connection – he knew of the helium plasma experiment that Tony Packard had given to one of his students. Helium would be just as effective as argon, and death would be just as pain-free. He resolved to find out more about how to access the gas. If only he were an experimental physicist, he would have an excuse to go into the laboratory. Then, out of the blue, he recalled a recent paper which described an experiment on the Abraham-Minkowski stress tensor. It struck him that, while further work would yield little new, it would provide an entrée into the laboratory.

Over the next few months, he became so absorbed in planning the details of his suicide that the tragedy of his situation was pushed into the background, almost, he reflected wryly, like air being displaced by helium. It was a bonus that the student happened to be using argon for the experiment as well as helium. That would be even more suitable, because helium would rise through the holes in the gas store's suspended ceiling and could leak out where the ubiquitous central heating ducts would almost certainly be routed through the walls above the suspended ceiling. Argon would fill the room from the bottom up, where it would be easier to contain with seals round the cracks.

After he had established his experiment in the laboratory, he went down one Saturday morning to check that it was locked, and then he had gone up to Reception to ask the janitor to borrow the key for the laboratory so he could retrieve a paper he said he had left there. There was a chance, of course, that the janitor would go down to open the door for him, but he wasn't known for his helpfulness, and, sure enough, the man just extracted the key from the key safe and handed it to him, telling him to sign for it. Then he had left the building by a side door at basement level, walked to South Street, where he waited while they made him a duplicate in the locksmith's and engraver's on the corner, and returned the key, signing it back in, all within the hour.

From time to time, the full horror of what he was about to do would surface, particularly when he was under stress, making his involuntary movements more pronounced, and he would become distracted and lose patience with people. On the whole, however, he took a bittersweet pleasure in tidying up the loose ends of his professional life as much as he could; on a personal level, there was little to do.

The incident with David Lane and his paper had been extraordinary. Until he had spoken to Miss Darling, he had not given the paper a second thought, only adding his name to lend it a little more credibility. After all, David had no track record in the subject, and he felt that he needed to sponsor it both as an author and as a member of the editorial board. It was only after Miss Darling had pointed out that the idea could be applied to the universe itself that he re-read the paper and found, to his astonishment, no flaw in her argument. Indeed, he had discovered more in the paper than he suspected even David Lane and Miss Darling were aware of, although David would inevitably discover it for himself if he were thorough in the preparation of the subsequent paper.

He was almost glad when his final weekend began. On the Saturday he posted two letters, both to the university: his mother had died ten years previously and he had no living relatives with whom he kept in contact. On Sunday, he dressed appropriately, putting on his pinstriped suit. He had chosen the Sunday evening as the time of his passing on. There was almost certain to be nobody around. Several times, on preceding Sunday evenings, he had gone down to confirm his prediction that nobody would be in the laboratory, always prepared with the excuse, if Mike had been inside when he tried the door, that he thought he had left a paper in the laboratory. Nobody was ever there on Sunday evenings.

So, now it was time. He surveyed his darkened office, gave it a little parting smile, opened the door, screwing his eyes against the light in the corridor, locked the room for the last time and walked lightly towards the stairs, his heart beginning to hammer. As expected, he encountered nobody. He let himself into the laboratory, switched on the light, locked the door from the inside, and crossed the floor to open the gas-cylinder storage room, switching on that light, too, verifying that at least one of the gauges of the two large argon cylinders was indicating full – the other was half full – and checking that the spanners were where he had seen them before, on the bench opposite the cylinders. Then he went back to the laboratory door, switched off the lights and entered the gas store room, closing the door behind him. He took from his pocket the four reels of black insulating tape that he had purchased from the iron monger's, switched off the store-room light and opened the window blind. When his eyes had dark-adapted sufficiently to use the minimal level of light filtering in from the quadrangle to see what he was doing, he began to seal round the edges of the window to make it gas-tight.

When he had completely sealed off the window, he closed the blind again and switched the light back on. He then turned to the door and sealed all around it. Next, he used one of the spanners to undo the nuts connecting the hoses to the low-pressure side of the gas regulators on each of the two argon cylinders and opened the valves fully. The gas escaped with an audible hiss.

He sat down on the floor, wedged in the corner between the door and the wall, the door hinge at his back, his legs straight out on the floor, bisecting the right angle of the corner, and asked God to forgive him.

After a couple of minutes had passed, he began to feel dizzy, and noted with a clinical detachment a numbness in his face and at the back of his throat. His vision began to blacken inwards from the edges. As he plunged down the tunnel, his oxygen-starved neurons began to fire for the very last time, signals ricocheting around his brain, hundreds of millions of stimulated neurons firing their charges in a final chain reaction, a last-in-a-lifetime firework spectacular, resurrecting unbidden images from long-forgotten alleyways in his neural labyrinth. All too soon the storm was spent, like the match girl's final blaze, but not before the dying embers of his self-awareness saw his mother, just as she had been when he was a child, before his world lost

its innocence, smiling and beckoning him onwards, towards
the end of the tunnel and into the light.

CHAPTER 16

Monday 29 November

Lucy guessed that she was considerably less stimulated than her fellow first-year students appeared to be by the sight of the three unattended police cars, two of them with their blue lights still flashing silently, drawn up untidily on the forecourt of the Physics Building. They filed past the cars on their way to lunch – many of them headed, like her, to New Hall. The chatter around her was speculative, the most credible of the rumours being that somebody had died in the building during the morning, although a close second was that the police had arrested a student suspected of being involved in terrorist activities. Lucy did not share their voyeuristic excitement, probably, she reflected, because, through David, she had grown to identify with the department and its people, and so its misfortunes were her setbacks, too.

David phoned her just as she was entering the New Hall foyer, and so she sat down outside the dining room to concentrate better.

'Lucy, are you all right?' were his first words.

'Why, what's wrong, David?' she asked, suddenly concerned.

'You've seen the police here? It's Jeremy Wilgoss. He's committed suicide.'

'Oh, God!' She paused. 'Are *you* all right David?'

'It's been bad, but we're coming through it now. I wanted to tell you before you heard it on the grapevine.'

'Thank you, David, that was thoughtful of you. Would you like to meet so you can talk about it?'

'Yes, I'd like that very much. Do you think I could? When would you have time?'

'I've got a physics practical this afternoon. How about tonight, after dinner, say half past six?' She would be able to catch up with her work on Tuesday and Wednesday. Anyway, it was nice that he had called her, that he wanted to talk with her.

'Thank you, Lucy – that would be great. Can I come for you at New Hall and we'll go for a drink? I know I'll need one to dull the memory of today!'

'I'll look forward to that. See you in the foyer.'

Over lunch, she began to feel less cheerful as the pleasure of David contacting her and the anticipation of seeing him so unexpectedly began to be crowded out by his actual message. She was at first sobered and then appalled by the knowledge that a person she had been talking to just over two weeks ago was now dead. Had he known, even as he was speaking to her, that he was going to kill himself? How could anyone know they were going to kill themselves and not give it away in what they said and how they looked and behaved? She tried to imagine the agony in the mind of someone who had decided to kill themselves, but she couldn't. She attempted to reconstruct their conversation and then a frightening thought hit her – did his abrupt departure from the couch have anything to do with his suicide? Had she said something that had triggered it? Promptly, she lost her appetite, stood up and carried her tray with her half-eaten meal over to the stack.

The afternoon passed slowly. She felt quite preoccupied and ill at ease during her practical class, in which she and another girl were trying to verify Stefan's law using an incandescent hundred-watt light bulb. She knew that the experiment must be straightforward, but she couldn't sort it out in her mind, and, to her shame, she ended up being largely a passenger throughout the session. At last, thanks to her partner, they were able to draw a straight-line graph

229

with the required slope and, at half past four, she left the teaching laboratory and headed towards the exit of the building.

She had only covered a few steps when she was intercepted by Paul Evans, who must have been waiting for her. Whatever could he want? She had felt a little shamefaced at their last parting, both David and herself having been effectively chided by him for wanting to hack into Wilgoss's emails.

'Lucy, hi! Got a moment?'

'Hi, Paul. What is it?'

'I just wanted your reaction to Professor Wilgoss's death.'

'Sorry, I'm not sure what you mean?' she said, continuing to walk slowly towards the front vestibule. 'It's dreadful, isn't it?'

'Oh, it's terrible. Have the police interviewed you, yet?'

She stopped in the middle of the corridor, her bowels liquefying. 'Why should they interview me?' she asked quickly, her voice sounding strained, even to her own ears.

'Well, with him trying to steal your paper and then dying just after.' He forced a little laugh. 'It looks suspicious, even to me, and *I* know you're innocent!'

'I don't know what you mean!' she protested. *This was awful.* 'Anyway, he didn't steal it, he just tried to block it, and David saw him about it and he unblocked it. End of story.'

'Oh, no, Lucy. I'm afraid that's not how it works. I mean, look at it from the police's point of view – to them, it's a pretty good motive for killing someone, especially for such an important paper.'

'He wasn't killed – it was suicide!' Lucy cried out, shocked to the core.

'Ah, was it suicide? Well, you already know more than the police, then. They've not said anything.'

230

'Look, I thought you were our friend. I'm sorry, I've nothing more to say. No comment!'

'I'm sorry, Lucy – I *am* on your side. Believe me, you really need someone who knows how these things work!'

For a moment, Lucy was tempted by his words. They really could do with a friend on the inside if that was how Wilgoss's suicide was going to be interpreted. On the other hand, Paul was first and foremost looking for a story.

'OK, Paul, but you have to understand I'm not saying anything further. In fact, if you ever use any of this in *Stat*, I'll deny it.'

'About your paper in the *International Journal of Natural Philosophy*? Oh, yes, I made a note of it. Don't worry, I wouldn't use anything without corroboration – I know someone in the journal's publisher – I'm sure they can verify that Professor Wilgoss tried to block your paper, and then none of this would be attributable to yourself.'

'Goodbye, Paul,' Lucy said firmly and resumed walking quickly to the exit. She could hear him following half-heartedly, and so she speeded up when she turned the corner. At the last moment, just before the exit, she took the stairs on impulse and ran up towards David's room.

Please be there.

As she approached his office, she could hear voices from behind the closed door, and then, spotting cloakrooms at the far end of the corridor, made a comfort-stop detour before returning and knocking on the door. David opened it and she was so relieved to see his delight at seeing her that she could have hugged him there at the door.

'Lucy, this *is* a surprise! Come on in. You remember Mike from the party? We share the office.'

Mike turned round in his chair and waved in greeting. 'Hi, Lucy. I'm sorry I'm in your way – the police have commandeered my laboratory, probably till this evening, they say.'

'Oh no, it's fine, Mike,' said Lucy. 'I just wondered if David fancied going out earlier – like now?'

'Yes, that would be great,' said David. 'I've not been able to concentrate on work anyway. Mike – will you be OK on your own?'

'No worries,' said Mike, adding that he would leave soon too, as his girlfriend would be coming home shortly.

David shrugged on his anorak and left with Lucy, the two of them walking side-by-side back to the stairs and out of the building.

They decided to go to the West Gate Bar again. 'I can't believe it's only, what, seven weeks since we were there,' said Lucy. 'So much seems to have happened since then that you kind of feel it must have been three or four months ago!'

'Perception of time's funny, isn't it? We spent five minutes waiting in the corridor this morning for the police and it felt like an hour, I swear. Lucy, it's very kind of you to make time for me so much earlier.'

Lucy thought she would wait until they were seated in the pub before she told him about Paul Evans' part in the decision. Instead, she let David talk about the events of the day. She was shocked to hear, for the first time, that he and Mike had discovered Wilgoss's body.

'Oh, David. No wonder you wanted to talk about it.' She wondered about asking how he knew it was suicide, but thought better of it and, anyway, they had arrived at the pub entrance.

They sat in the same part of the bar, next to the alcove that they had occupied after Wilgoss's lecture. It occurred to her that it might not have been the best choice of pub, for that reason, but David made no reference to the circumstances of their previous meal, and, having ordered from the bar, continued to talk about what had happened during the day.

'My treat, tonight, Lucy, since I'm bending your ear,' he said.

'Sorry, David, you know the rules! We go Dutch.'

'I thought you'd say that. Anyway, the wine's definitely on me,' and he filled their glasses from the whole bottle of Merlot he had brought back from the bar, taking a large gulp from his own glass before continuing.

'It happened very quickly. Mike had seen it was Wilgoss before he passed out. I thought it was from shock, and I felt faint myself, but Mike said he saw the hoses from the argon cylinders had been disconnected. He reckons not all the argon had leaked away and it came pouring out when he opened the door – that's what made him faint.'

'So that's how you know he committed suicide – the argon cylinder?'

'Yes, but the clincher was that the door had been taped shut, and you could see it was from the inside. I suppose that was to seal it for leaks.'

'So, presumably the police could see that, too, as soon as they arrived?'

'You'd think so, wouldn't you? But, no, they have protocols. The police who were first on the scene didn't do anything except keep us and everybody else out. And they segregated Mike and me and Tony Packard, too, so we couldn't compare notes, I suppose.'

'But they *have* said it's suicide now, haven't they?'

'No, actually, not as far as I know; at least, nobody's told me. But it's just a matter of time.'

'Oh, I very much hope so,' said Lucy.

'I have to admit, before the investigation team arrived to interview us, I was worried they would find evidence of his emails about the paper on his computer and think I might have pressurised him to commit suicide.'

Lucy recalled her own anxiety along the same lines, but she held her peace for the moment.

'Then something extraordinary happened,' he said, pausing to take another long draught of wine, Lucy keeping him company.

'I was just sitting there, waiting with my minder, my policeman, and then I couldn't sit there doing nothing any longer, so I asked if I could go down and collect my mail.'

He waited while a waitress brought their order – two steak ciabattas – the same selection that they had chosen on their first visit.

'So he said, no problem, and we went down to reception together.'

'He followed you everywhere?'

'Yes, he did.'

She wondered if that included going to the loo, but decided she wouldn't ask. 'Sorry, go on,' she said.

'Well, Ann, the receptionist, gave me my mail from my pigeonhole, just like she does every morning, and I shuffled through it as I went up the stairs with my minder, and then I saw it – a plain, white envelope with a first-class stamp and the address written by hand in black ink. Well, that's pretty unusual, as you can imagine.'

'Yes, I suppose your mail's all brown typed envelopes and colour catalogues.'

'You've got the picture. Anyway, the most startling thing about it – not to say the most distressing thing – was that it was from Wilgoss!'

'Oh, God. And you had just seen him dead!'

'Exactly! Well, of course, I recognised the writing but the policeman had no idea there was anything amiss. So I said to him that I had to go to the photocopying room on the way back and he was happy with that. So we changed tack and headed downstairs again to the photocopying room and I quite calmly slit open the letter with my finger and took it out of the envelope so that when we got to the photocopier, I was able to hold all the envelopes in my hand while copying the letter – it was only a single sheet – and the

234

policeman didn't even bother to look. I copied a couple of other things as well, just to make it look like that's what I always did.'

'I don't understand – why did you want to copy the letter?'

'I thought it might be a suicide note. If it was, then I wanted a copy, because I'd have to declare it and the police would take it away from me. So I shoved the copied page in with the rest of the envelopes and carried on upstairs. When we were back in the office, I slit the rest of the envelopes open and put all the contents on my desk, and casually put some of the stuff in my drawer, including the copied page.'

'And you hadn't read it yet?'

'No, but then, when I saw the policeman wasn't looking at me directly, I read the original letter. It didn't take long.'

'And was it a suicide note?'

'Well, yes, I suppose it was. It was dated Saturday, so he must have posted it then, knowing I'd get it after he had died. Very calculating of him. Anyway, this is what he said, more or less. I've read it so often – the copy, I mean – that I know it off by heart. It said that by the time I had read it, I'd know he'd taken his own life. He hoped I wouldn't entertain any thoughts that our recent discussions regarding our paper could have influenced him in any way in taking his life. He said that his motivation was entirely for medical reasons, which he had detailed in another letter to Professor Oakhill. He ended up saying – get this – "May God guide you in your deliberations over any subsequent paper". Creepy, huh?'

Lucy said nothing, trying to imagine how Wilgoss must have felt writing the letter.

'Did you tell the policeman?' she asked, finally.

'Yes, I made out as though I'd just opened the letter, of course. I guess they'd really have wanted me to give it to them unopened, so they could check that only Wilgoss's fingerprints were on the inside, but I badly wanted to know what it said.'

'Did he take it from you?'

'Yes, he put on gloves and put it in a plastic bag, but he didn't say I shouldn't have opened it. Later, when the investigating detective interviewed me, he did ask why I'd opened it, but I said it was an automatic reflex to open your mail, and he seemed satisfied with that.'

'And to think it was *me* you accused of having the criminal mind!' said Lucy.

'I saw John Oakhill as he was dashing out to catch a train after his interview with the detectives, but he stopped long enough to tell me he'd received a letter, just like Wilgoss said. The letter said he had Huntington's disease and had decided to kill himself.'

David went on to tell Lucy about Huntington's. Oakhill hadn't had time to explain, but David had read it up on the internet after he had been interviewed.

'Did he have any family?' Lucy asked.

'No, there was nobody, as far as I know. He was a bit of a loner.'

'I wonder if that was deliberate? Not having children, I mean.'

'Maybe. We'll never know.'

They talked for a long time about how it must feel to have an evens chance of developing the disease, bringing the meal to a rather sombre close. When Lucy felt David had talked himself out, she started to tell him, with a heavy heart, about her meeting with Paul.

'I met Paul Evans just before I saw you this afternoon,' she said.

'Oh, yes, I forgot to tell you – he came to see me too.'

'What did he want?'

'Oh, for me to dish the dirt on Wilgoss. He wanted to know what the connection was between us coming to see him about the paper and the suicide.'

'What did you tell him?'

'Nothing. I said *no comment* when he mentioned the suicide. Actually, he called it a death.'

'Oh! I'm afraid he tricked me, scared me, into saying a little more.'

'Oh, poor Lucy! What did you tell him?'

'I told him Wilgoss tried to block the paper but that he later unblocked it. I wouldn't have said that except that Paul said Wilgoss was trying to steal it, so I put him right.'

'Well, no matter,' said David. 'Paul's a slippery character.'

'He said he could get corroboration about the paper from a contact on the *International Journal of Natural Philosophy*.'

'Well, in that case, we're in the clear,' said David. 'I'll probably tell John Oakhill just to forewarn him in case Paul uses it for *Stat*. Don't worry about it.'

Lucy felt almost overcome with relief – David had such a reassuring manner, like a good doctor. 'David, you're so kind to me,' she said throatily. 'I was the one who was supposed to listen to you telling me about the awful things that happened today and it's you who ends up consoling *me*!'

Without warning, she felt tears welling up in her eyes, and in a dissociated way, realised that drinking three glasses of wine probably had something to do with it. At that moment, thankfully diverting David's attention, the waitress appeared at their table, cleared away their plates and asked them if they wanted to see the pudding menu.

David looked across at Lucy, now back in possession of herself. 'I think we could do with something stodgy to soak up the alcohol,' he said good-naturedly. 'Do you fancy anything?'

You. 'Sticky toffee pudding with ice-cream sounds good,' she said.

They decided not to have coffee at the end of the meal but instead to walk back to New Hall where Lucy would make some. She felt as though she were being given a second chance; she was interested to note that she felt none of the inhibitions that had prompted her earlier prevarications about taking him up for coffee. Probably the biggest change was that their relationship had taken a new direction during the course of the day: they had supported each other against the current of the day's adversities, helped each other to keep their footing until the threatening flow had abated. Now they were officially more than just professional colleagues.

Arriving at the hall, they took the lift to the third floor, and walked along the corridors to her room at the east end of the building. 'Welcome to my home,' she said, turning to greet him, executing a half pirouette, arm extended to include the whole room in her best untutored port-de-bras – *all this is mine*.

David smiled. It seemed impossible to her that he was actually here in her room, looking even taller against her furniture and fittings, normally scaled by her subconscious eye from her own, shorter perspective.

He was immediately drawn to the Monet prints on her wall. 'Ah! *Impression, Sunrise*. Oh, and *Sunrise at Sea*. I love those paintings.'

Lucy was thrilled. 'You know them! I love them too!' She laughed. 'Obviously,' she added, 'or I wouldn't have them. This is my favourite, though.'

David looked where she was pointing. 'Ah, yes,' he said, '*The Magpie*.'

'I love the way it brings back the best memories of winter,' she said, 'the hush when everything is covered with a thick blanket of snow, hiding the dirt, the imperfections of daily life. I love the way the sun catches the edges of puffballs of snow like the golden-yellow lining of a cloud, how the wall's shadow on the snow is actually pale violet – it doesn't come out well in this print – so that it must be reflecting the sky above, that you can't see in the picture.'

'So you've seen the original painting? Where is it?'

'Yes, in the Musée d'Orsay in Paris. We went to Paris for a family holiday. I'd already seen the picture in a book – I've still got it at home – and so I pestered Mum and Dad to take me to the Musée d'Orsay because I knew that's where it was kept. It's thrilling to see a painting in real life that you already know so well from a book, and to know this is the actual canvas painted by Monet himself.' She went over to the coffee maker. 'Have you ever had that feeling?'

'Yes, I have, in fact. In the National Gallery, when I first saw *The Fighting Temeraire*. Same as you, I had a picture of that in a book that I used to gaze at for hours on end when I was a boy.'

They discussed their common interest in impressionist paintings while Lucy prepared their drinks and poured out a coffee for each of them, David sitting with his back to the window and Lucy leaning with her elbow on the desk, her chair turned to face him. He cocked his head to the side to read the titles on the spines of the books on the wall shelf.

'Hah! I remember those,' he said, pointing to the first-year physics text-books. 'I've still got mine, for all of the years. You're welcome to have any of them you like. Promise you'll ask me before you buy any?'

'But don't you ever refer to them, now?'

'Not the undergraduate ones, to be honest. It's much quicker just to use the internet if you need reminding. But you need the books in the first place so you know what's bona fide information when you start out. Oh, good – I can see fiction! Excellent!'

'*Pride and Prejudice*. I think it's probably my favourite book of all time. I've probably read it more often than any other.'

'What – you would choose chick lit over science fiction?'

She was about to protest when David anticipated her, laughing. 'Just kidding! What's your favourite science fiction book?'

'My favourite science fiction *story*,' replied Lucy carefully, 'has to be *The Last Question* by Isaac Asimov. Actually, it was Asimov's favourite, too – at least, the one he liked best of his own work, which was probably the same thing.'

'Yes, that's a brilliant story. Every time I read it – it's only ten, fifteen pages, isn't it? – I still get a tingle down my spine. Funny enough, if you think about it, it's not a million miles from what we're trying to do ourselves. This is excellent coffee, by the way.'

'Good, I'm glad you like it. How is it like what we're trying to do?'

'Well, aside from the fact that it's mildly self-referential – you know, the computer generates the universe which generates the computer – what I was really thinking about is that the *last* question has two meanings. You remember, in the story, at the end of time, at the end of the universe, when even space and time no longer exist, the ultimate descendant of the Multivac computer labours for a timeless interval to solve the question of whether entropy can be reversed. Then, when it comes up with the answer, it creates the universe – *Let there be light – and there was light!* – beat that for an ending to a story! So the last question was literally the last one to be asked and solved. But Asimov also means it in the sense of it being the ultimate question – the answer to which is life, the universe and everything. And, essentially, that's what we're doing by working on the higher mathematical system, the ultimate system, the God System.'

David's retelling of the story reminded Lucy of a niggling question that had bothered her earlier and she had

240

later forgotten. 'I was always a little uncomfortable with one aspect of the story,' she said. 'At the end of the story the entropy of the universe has increased to a maximum, of course. But that means that the information in the universe has increased to its maximum also. Then all that information just disappears when the universe begins again, because, at the very beginning, the information content must be relatively tiny.'

'Yes, you're right, our universe started with just a few bits of information, and, at the moment, it has something like ten to the power of one hundred and twenty bits. So it does beg a question about what happens to the information.'

'Yes, but that's not exactly what's bothering me,' said Lucy. 'I was thinking more about our own paper. If the underlying mathematical system for our universe is essentially simple, then where does all the information come from, and how can it keep growing with time?'

'Very good question, as always! One way to look at it is to think of the Mandelbrot picture. That seems to get more and more complex the closer you examine it. But it's still very simple – it's still all produced with an equation. The apparent complexity increases the more you apply the equation – the more you iterate it – but it's all completely predictable. Now, of course, the universe is very different from the Mandelbrot set, but the principle's the same.'

He took a drink from his cup. 'With at least one important exception,' he added. 'Quantum uncertainty! The truly random nature of quantum phenomena is responsible for much of the increasing information content of our universe.'

'Well, if that's the case,' said Lucy, 'then there *is* a problem – the information content of the universe really *is* greater than in the generating equations.'

'Well, I suspect that the resolution to that is having all those parallel universes – when you add them together, all the different possible outcomes of any quantum interaction

241

– the sum over their individual histories, if you like – largely cancel out, leaving quite a bland, predictable multiverse when you regard it as a whole. But we needn't depend on there being parallel universes – the point is that the higher mathematical system will determine which quantum interactions actually take place in our universe – and all the other parallel universes, assuming they exist.'

He drained his coffee. 'Another way to look at it,' he continued, 'would be to see that, in the case of quantum events, no experiment in this universe can predict the exact outcome with a hundred percent certainty. So the outcome is undecidable, at least in this universe. But in a higher system – it may be like the solution to the undecidable Gödel sentence – there may be a proof: each outcome may be completely deducible.'

'But surely that would mean that quantum events are not random after all? I thought that experiments had proved that they are?

'Oh, quantum stuff's random all right, at least in our universe. There's no way that anything in this universe can predict the outcome of a particular quantum event in this universe. But, to take another simple example, if you write down the digits of Euler's number, e, they're random, at least as far as anyone can tell – except for the fact that you can generate every one of the digits predictably from the formula for e! I'm not saying there's a pseudo-random-number generator in a higher system behind quantum mechanics – far from it – but I am saying that the apparent increase in the information content of the universe may only be apparent in our universe. Seen from the higher level, all events are provable. It only looks like true, unpredictable information to us lowly beings – we who are, ourselves, stuck in our universe as the products of the mathematical system of our universe.'

Lucy had said little during this exchange as she was finding it a little difficult to follow. Evelyn Waugh's phrase

242

unused to wine popped into her head. 'He published quite a few stories about Multivac, Asimov did,' she said. '*All the Troubles of the World* was kind of self-referential too.'

'Oh, yes – was it? How?'

'Well, if you remember, in that story, Multivac has become so complex that it's become sentient, though nobody knows this yet. It has been given all the problems of humankind to sort out, and it wants an end to that burden, so it arranges for a teenaged boy to switch it off. But Multivac still has the innocence of a young child and so it reports to government officials that there is a high probability that someone is going to commit a murder, though it doesn't name the intended victim. The officials don't realise until it's almost too late that the computer is anticipating its own murder – the very murder that it has been orchestrating through the boy.'

'Now there's an appropriate question for today,' said David. 'Can a computer commit suicide? For that matter, would it be morally reprehensible – not to say murder – to switch off permanently a computer that was sentient?'

'Which brings us back full-circle to self-awareness and the meaning of a soul,' said Lucy. She stretched, yawning, suddenly tired, and David rose out of his chair.

'Sorry, Lucy,' he said. 'You must be exhausted. It's been a really long day for both of us.'

Lucy's hand shot to her mouth. 'Oh, no, David, sorry. I didn't mean to do that.'

'No, really, I'm totally whacked, myself – I need to get home and get twelve hours' sleep. And you have to be recharged for tomorrow's lectures.'

'Yes, I suppose you're right. Let me show you downstairs.'

'Oh, no, that's OK, I'll find my way.'

'Don't count on it,' she said, 'this is a real maze,' and she led the way out of her room, locked the door behind them, and accompanied David in the lift down to the foyer.

243

The sliding glass doors detected them and opened as they approached the entrance.

David turned to face Lucy. 'It's been an extraordinary day, Lucy,' he said, 'and you've helped me get through it. Thank you.'

'I was just thinking how you helped *me*,' she laughed, 'especially after meeting that Paul Evans. It was a good idea to go to the bar – it was just what I needed.'

'Oh, I thought the bar was your idea.'

The doors decided they'd been open long enough and so they started to close again slowly, only to spring back at the approach of two girls returning from an evening out.

'Well, I really must get going,' said David. 'Thank you ever so much for the coffee and everything, Lucy. Sleep well.'

Lucy waved as he went out into the night. 'Goodbye,' she said softly into the darkness. *I love you.*

CHAPTER 17

Thursday 2 December – Saturday 4 December

It was not until Thursday morning, three days after Jeremy Wilgoss's death, that David felt that he might be able to do some useful work on his thesis. Over the past couple of days, time and again he had read over the last few pages that he had written before Wilgoss had died, but it was as alien as reading somebody else's paper: he was unable to recover his frame of mind when writing the paragraphs, and so he couldn't quite recall his ideas for the next section of the work. This morning, however, as he had walked gingerly to the Physics Building, taking care with his feet on pavements made slippery by an unexpected early-December fall of snow during the night, his plans had all come flooding back to him, in the way that an elusive name will pop up unannounced when the mental search has been abandoned and attention has been reluctantly directed back to the humdrum processes of daily life.

He was working alone in the office at eleven o'clock – Mike had taken himself off to the laboratory, which the police had only released the day before – and he was just contemplating going down to see if Mike was ready for morning coffee when there was a knock on the door.

'Come in,' he called, and John Oakhill entered the room, reminding David of the visit from Wilgoss the previous week – goodness – was it really only just a week?

David offered Mike's chair and Oakhill sat down heavily. 'I must be getting old,' he said. 'First, David, I must apologise for not seeing you sooner. As you know, I had to go to London on Monday as soon as the police were finished with me and I got back late last night. How are you bearing up?'

'Oh, you know – not too bad, considering.' In fact, the image of Wilgoss propped up in the corner with his eyes staring open had been a constant accompaniment to all of his waking thoughts. He had only once before seen a dead body – that of his mother in hospital after her devastating brain haemorrhage. The doctors had kept her body functioning on the ventilator until he had arrived, but they had told him she was brain dead. When he had said his goodbyes, they had switched off the ventilator and he and his father had stayed while some primordial neurons still functioning in her brain attempted to breathe sporadically for her, the interval between each struggled breath increasing inexorably, her eyes disconcertingly opening, seemingly straining with the effort, as though she were conscious. Finally, she had stopped, like the ventilator, eyes wide open, like Wilgoss.

Oakhill nodded and stared out of the window, gazing across the courtyard, contemplating, almost as though he could read behind David's words, giving him time to say more if he wanted to. Eventually he looked back at David. 'The Vice-Chancellor phoned this morning. The Procurator Fiscal has declared it a suicide and the investigation's closed.'

'That was quick. I thought there had to be an inquest or something.'

'Aye, so did I, but apparently that's in England. It seems that the autopsy confirmed death by asphyxiation, forensics are satisfied, and, what with our two letters – we'll get them back soon, I gather – the Procurator Fiscal decided that no further investigation is necessary: he's ruled it a suicide.'

'I must say, that's a relief,' said David. Now would be a good time to let Oakhill know that Paul Evans was likely to say something about their paper in *Stat* when it came out the following day.

'John, there's something I wanted to talk to you about,' he said.

'Ah, yes, that was the main reason I came to see you.'

David was taken aback. 'Sorry?'

'You must be wondering what's going to happen now that you're without a supervisor.'

Actually, it had not occurred to him until that moment.

'Well, of course,' Oakhill continued, 'I'll be very happy to take you on. If that would suit you, naturally.'

'Yes, thank you, John. I'd like that very much; that would be very kind of you.'

'Oh, not at all,' he said, waving it aside. He took out his diary and consulted it, a small A6 book with a page per week. 'For starters, maybe we could agree on a time for an hour's meeting, say, so you can give me an idea of what you've been doing.'

They arranged a meeting that was convenient for Oakhill in the following week, David being totally flexible with virtually no commitments.

'There's something else I need to tell you, John. It's about Jeremy's death – it'll be reported in tomorrow's *Stat*.'

'Aye, I've no doubt it will. What of it?'

'Well, it might say that Jeremy tried to stop us from publishing a paper. We talked to Paul Evans.' There was no need to elaborate on the name – it was sufficiently notorious among staff who had been at the university for any length of time.

Oakhill became serious. 'What do you mean? Who's *we*?'

'A physics student and I – Lucy Darling, she was at your party, remember? – we've written a mathematical paper that indicates that any mathematical system that's self-consistent and sufficiently complex must be a subset of a higher system that can't be accessed logically from that system.'

Oakhill raised a hand, halting him. 'Steady, steady. You'll need to run that by me again in a moment, but what's this got to do with Jeremy?'

'Well, he offered to smooth its path through the *International Journal of Natural Philosophy*. He was on the editorial board.'

Oakhill nodded, yes, he knew Wilgoss was on the board.

'Then he came and said both referees had rejected it. I took him at his word until I asked for a copy of the reports, so we could maybe improve the paper, you know, and he just refused to give me them. Then I wondered if he'd maybe altered the paper in some way and so I went to Paul Evans to see if he could help – you know, using his journalistic methods and contacts.'

'Aye, I know very well. And you should have known better. If you had only come to me in the first place, I could have helped. Colin Howarth, the editor of *Int J Nat Phil*, is a good friend – we were both at King's College, Cambridge, as undergraduates.'

David felt his cheeks burn. Of all the lecturers at the university, John Oakhill was far and away the one he most respected and the one he would most wish to please. To be admonished by him, however mildly, made him feel so wretched, as though he had let down both himself and Oakhill.

'Och, no matter,' Oakhill added, clearly sensing David's discomfort. 'It isn't the end of the world.'

David was grateful for the lifeline, but he hadn't finished. 'I know that was idiotic of me. As it happened, Paul didn't help us, but I'm afraid I was stupid and told him that the paper held a clue to the origin of the universe. He also got out of us that Jeremy had withdrawn the paper, and he was going to check with his sources on the journal.'

Oakhill had to smile. 'The origin of the universe, eh? I can just see the headlines now! You were right to let me know about this. I'll call the Vice-Chancellor so he's forewarned, but don't worry, he's not going to be annoyed, at least, not at you personally. But how do you know Jeremy withdrew the paper?'

'I checked with the journal. Jeremy withdrew our paper after the referees had both recommended publication. I went to him about it and he said that the paper would be quoted in the popular press as a proof that God exists, and that wasn't what religion was about, and that's why he withdrew it.'

'So what happened to the paper?'

'Well, after I confronted him, he contacted the journal and asked them to publish it after all, but he took his name off it.'

'What's at the root of the paper?'

'It was Lucy's idea. She reckoned that the undecidability of Gödel's sentence in any mathematical system implied a higher system in which the sentence could be proved. Between ourselves, we call it the Inverse Gödel Theorem. I supplied the formal mathematics.'

Oakhill raised his hand again and held it there, eyes closed, for twenty seconds before lowering it and looking back at David.

'If your paper does what it says on the tin, then it contains dynamite,' he said, finally. 'What's your next move?'

'Well, we're working on a follow-up paper to apply it to whatever mathematical system underpins the universe. It's still work in progress.'

Oakhill looked solemnly at David and said 'Let me suggest that, when you write your second paper, you don't call the theorem in your first paper the *Inverse Gödel Theorem*. Then, who knows, when people refer to it in future, they'll maybe call it the *Lane-Darling* theorem.' Then he winked.

'It's eleven-thirty – time for your early-morning coffee,' he said, and, with that, he rose from the chair and made for the door.

'Thank you, John,' said David, his voice full of gratitude, and stood up himself to winkle Mike out from his laboratory.

*

Two days later, on Saturday morning, David stood looking over the railings into the vast grounds of the ancient, ruined cathedral. A sound-deadening haar, a sea fog, had rolled inland overnight, almost unknown for the time of year, but then, he reflected, so had been the early fall of snow which had vanished, literally evaporated, sublimated, on Friday with the crossing of a warm front. In the foreground, a crowd of ghostly white and grey gravestones huddled beneath the dark skeletal branches of winter-bare trees, silhouetted against the luminiferous haze of the mist, back-lit by the rays of a weak, December sun hiding beyond the fogbank. A hundred yards into the greying distance, the twin turrets of the solitary eastern gable loomed over the far end of what used to be the nave, now a flat, immaculately cut lawn peppered with yet more gravestones and the time-worn stubs of massive ruined pillars, long since vanished, plundered for their building stone.

A movement caught the corner of his eye and he turned to his right, expecting to see Lucy appear round the corner from South Street, but it turned out to be a woman walking her spaniel, the black dog-waste bag which she carried with such civic pride swinging from her hand like a scrotal sack.

'Hi, David,' Lucy cried out behind him, waving as he wheeled round. She had come from the corner of North Street instead.

'Morning, Lucy. Not quite what I imagined, I'm afraid,' he said, raising his arm high behind him to indicate the fog.

'Oh, that's OK, we can see well enough. At least it isn't raining.'

'Yes, but I wanted to take us up Canon's Tower. The view's stunning on a good day.'

'Where's Canon's Tower?'

David couldn't help laughing. 'Exactly! In fact, you can just make it out to the right of the eastern façade with the twin pinnacles – do you see that tall, ghostly shadow?'

'Oh, I see what you mean. Yes, you wouldn't see from up there today, would you? No matter, I'd still like to explore the cathedral grounds, if you're up for it.'

'Sure. In fact the grounds are free – there's a charge for the tower. We just go in that gate,' he said, nodding to the entrance ten yards further along.

'So what's the history?'

'Rats, I knew I should have mugged up on it! I'm afraid all I know is that it's about eight hundred years old and time hasn't been kind to it. Various bits were blown down in storms or destroyed by fire or simply broken up for building material until all that's left is what you see here.'

'It's still impressive, what's still standing, I mean. It must have been an enormous building in its time.'

'Yes, I believe it was some sort of record-breaker. You know, when Christos – you remember Christos? – well, when he first arrived here from Greece, the taxi dropped him off there' – he pointed to the old stone building across the road from them – 'and he didn't realise that those were his lodgings, where some of the postgrads live: he took one look at the cathedral, thought that he would have to stay in these ruins, and his heart sank.'

Lucy laughed. 'You're kidding me!'

'No, that's what he told me, at any rate. He'd heard that it was an ancient university!'

They turned in through the open gates of the cathedral grounds, passed under the archway of the western gable and headed eastward down the centre of what must have been a truly magnificent, awe-commanding nave.

'So, how have you been?' she asked.

David was touched by the tenderness of her tone. 'I'm pleased to say I'm getting back to my usual self. Thank you for asking. How about you?'

'I keep trying to imagine how it must have felt for poor Wilgoss to take his own life. And then there was that awful *Stat* article yesterday.'

As expected, the front page had been devoted entirely to the death. While there was nothing factually wrong, which was, in itself, surprising, Paul Evans had nonetheless put an unpleasant spin on the story. He had written that Professor Wilgoss had taken his own life after he had been discovered trying to hide a key paper which would ultimately unlock "the central mystery of the universe". It could be read as though he had committed suicide directly as a result of David and Lucy catching him red-handed. He had reproduced the same picture of Lucy that had been used for the choking story. John Oakhill had come to see David again and told him the Vice-Chancellor was drafting a letter for the next *Stat*, putting the facts into perspective, and David had called Lucy that evening to reassure her. It was then that they had arranged to meet at the cathedral.

'Well, hopefully the Vice-Chancellor's letter will put an end to it. It's not just embarrassing that Paul referred to our paper, it's actually quite annoying.' He stopped as something occurred to him.

'I've just remembered something else Wilgoss said to me the last time we met. He said that I might be surprised at what's in our paper. That was a curious thing to say, wasn't it?'

'What did he mean by it?'

'That's just it, I don't know. I wonder if there's something in there that we've missed?'

'Enough to put him over the edge?' Lucy asked, apprehensively.

'Not for a moment, Lucy. It was to avoid the final stages of Huntington's disease. But I must have another look at our paper soon, to see if I can figure out what he meant.'

They were half-way down the nave, just before it was crossed by the transept, when they came to a well, protected by a circular, black cast-iron grille a few feet in diameter.

'Now why would there be a well in the middle of the cathedral?' asked Lucy.

'I think it was for holy water. Guess how deep?'

'No idea. Ten metres. So how deep is it?'

'No idea. Got a stone?'

They looked around for one but there was not the smallest pebble to be found.

'They keep it pretty tidy, don't they?' said Lucy.

'Either that or the visitors used them all up,' said David, extracting a two-pence piece from his pocket. Here, you do it.'

'Sure? OK, thank you.' She knelt down and reached over to the centre of the grille and dropped the coin.

'A thousand and one, a thousand and—' she counted, stopping as soon as she heard it hit the bottom.

She turned to David. 'No water! So that's one point eight seconds squared – call it three and a quarter – multiplied by ten – that's about thirty-two and divided by two – so it's about sixteen metres deep.'

'Which is about fifty feet – that sounds more impressive,' said David. 'Did you make a wish?'

'I didn't think you would be superstitious. I'm not!'

'So did you?'

She laughed. 'Yes, but before you ask, I'm not going to tell you what I wished for.'

David laughed with her. He knew what his wish would have been.

'Does this remind you of a certain tutorial question and a student's spectacular answer?' he said.

Lucy gave a shy laugh. 'It wasn't spectacular at all. In fact, I reckon I got off lightly, not having to work out the tedious equations.'

'That's just my point. You have a gift for seeing behind the equations to their underlying meaning. I believe too many physicists think in terms of equations without grasping the bigger picture of what the equations are representing.'

'In a way, that's ironic, isn't it? I mean, you say the universe can be reduced to a set of simple equations, rules, which generate the universe, and I believe you're right, but surely that's like saying that the fundamental thing is the equations rather than the bigger picture they represent.'

'No, I don't believe there's a contradiction in that' said David, speaking slowly, thinking carefully about what she was saying. 'The basic rules will be very simple – it's the unimaginably huge number of ways of combining them over and over again that generates the complexity – and usually our simple description of the picture is an approximation – a very useful approximation, granted – to what is actually happening.'

'How do you mean?' Lucy asked.

'Well,' – he sought around for an example – 'just take the coin falling into the well. You applied a simple equation to get the approximate depth. But, of course, in reality, the coin is spinning, the air surrounding it is turbulent and slowing it down, the force on it isn't exactly constant all the way down and so on and so on. The equation you used gives a useful but approximate picture. However, all the complexity can be described in terms of a few very simple relations between quarks, gluons and so on, or even in terms of strings or whatever is the most fundamental level.'

'Do you think it would be possible to write down the equations – the rules – of the meta-system – even the God System – in our universe, if the equations are ultimately simple?'

'Hah! What an idea! Well, for starters, of course, if it were possible, we would still be unable to prove that these were actually the equations of the meta-system or the God System, by the definition of a meta-system.'

'OK, yes, that's a given. So what do you think?'

'Well, provided that the God System isn't too large, then, yes, probably you could write them down. But, as I said, you'd never be able to prove that these were the equations of the higher system containing the mathematical system of our own universe.'

They had reached the eastern gable end of the ancient nave which now loomed above them, the details of the twin pinnacles lost in the haze, and so they turned back to the right, walking past the gravestones towards the cloister abutting the south wall of the cathedral.

'I've just thought of an analogy,' David went on. 'Suppose that the inhabitants of Flatland, who are, of course, well versed in the axioms of Euclidian planar geometry – suppose they ask this question. Why does the fifth postulate, as it's called, say that given a straight line and given a point not on the line, then only one straight line can be drawn through the point without intersecting the first straight line, no matter how far the lines are extended?'

'Oh, I see,' said Lucy. 'They might have chosen an axiom, a postulate, that said *no* such line can be drawn through the point rather than *one* such line. That would give them straight lines that actually curve round matter, like in general relativity in our own universe.'

'Yes, and if they had chosen an axiom that said more than one line can be drawn through the point, that would give them hyperbolic geometry which is another possibility for our universe.'

'So what you are saying,' said Lucy, 'is that more than one, maybe an infinite number of self-consistent mathematical systems representing possible meta-systems might be written down in our universe, but we'd have no way of proving which was correct – not in our universe, at any rate.'

'Exactly.'

'But I wonder if there might be some clue within the equations that you write down that would tell you that this just has to be right, even though you can't prove it? Like Watson and Crick just knew, when they worked out the structure of DNA, that it *must* be right, because they could see immediately how the molecule could replicate itself.'

'Good point. Imagine how exciting that would be!'

They had come to the cloister, entering through one of the series of archways in the boundary wall and following the perimeter path, deep in thought like the canons who had trodden the same ground centuries before them.

'Another well!' said Lucy. 'Your turn.'

'OK,' said David, fishing another coin from his pocket, leaning over and counting while it dropped. 'A thousand and one— ah! Only one second.'

'Only five metres then, fifteen, sixteen feet. Did you make a wish?'

'Sure I did, but you didn't tell me yours, so I'm not telling you mine!'

'Rotter!'

'This mist is chilling me to the bone – do you fancy going for a coffee?'

Lucy jumped at the idea and so they made their way back to the gate and headed a little way down South Street to Ferarri's café.

'Have you been here before?' David asked as they found a table, opening up their anoraks to let in the welcoming warmth.

'No, I depend totally on you as my guide to the town,' said Lucy, smiling. 'I know they're famous for their ice-cream, though.'

'It's delicious beyond description. But right now, I'd prefer one of their hot bacon baguettes. They're really tasty. Will you join me?'

'So what's the next step with the second paper?' Lucy asked when he returned from ordering the coffee and baguettes.

'Well, I had been thinking of an essentially descriptive paper rather than a mathematical one, but I'm going to re-read our first one to see if I can discover what Jeremy reckoned he saw in it. I feel that the pressure's on, now, because of the *Stat* article. You never know who might pick it up from the web – we need to keep ahead of the crowd.'

'David, if or when these two papers are published, will you feel in any way that there's nothing left to be done in physics – that the ideas in these papers are so fundamental that all the rest is just about mopping up the crumbs?'

'Oh my goodness, no! If anything, it makes physics even more exciting.'

'How do you make that out?'

'Because there are going to be puzzles in physics, in the universe, which are undecidable – we might guess which ones, but we can never be sure, by definition. The origin of our own universe might be one – was it a big bang, was it triggered by a clash of branes, is it cyclical, did it evolve through cycles of black holes, or big crunches and bounces or never-ending epochs – although, I suspect, we shall eventually find enough evidence for one of these, or perhaps, something quite different. And, as you were suggesting back there, if we try to guess at the mathematics of the God System, there may be some properties that shout out at you so you know it just *has* to be right – then we can work back and see what such a system would predict about the undecidable puzzles and about the mathematics

257

underpinning our universe and the multiverse. The strange things in quantum theory have to be near the top of the list. If we can make a guess at the form of the God System, then we might be able to start ruling in or ruling out some of the competing theories. Millennia of fun for physicists!'

'Good, I'm glad,' she said.

'What – were you worried about not having anything to do in future?'

'No, I meant I'm glad for you. You have such an enthusiasm – an infectious enthusiasm – and I wouldn't want to see you lose it.'

'Thank you, Lucy. That's really nice of you.' David was touched that Lucy showed such concern for his spirit, although he shouldn't have been surprised, thinking back, because she had shown similar thoughtfulness ever since Wilgoss's suicide. He desperately tried to think of a way to follow on while leaving himself a way to backtrack in case he had misinterpreted her solicitude, when the waitress brought their order, and the moment was gone.

They munched though their bacon baguettes in companionable silence and resumed their conversation over the coffee, David explaining the research work for his thesis in greater detail, and Lucy recounting for him some stories about the students that he knew by sight in the tutorials. He was pleased to hear that George had merely been a nuisance, and not her boyfriend, and was greatly amused to learn that, apparently, he had stopped pestering her after the question about the tunnel through the Earth.

In due course, they left the café and walked side-by-side down South Street, along Bell Street and into Market Street, David heading for his house and Lucy for New Hall. They parted at the corner of Spey Street, David stopping as he had before, watching Lucy, and, sure enough, after she had gone a few yards she turned and waved to him. This time he waved back.

CHAPTER 18

Monday 6 December – Saturday 11 December

As Lucy walked away from the Physics Building at the end of the morning lectures on the following Monday, she was in a contemplative mood, only dimly aware of the freezing rain that spattered down from a slate-grey sky onto the taught nylon of the collapsible umbrella that she had taken out of her anorak pocket, erecting it automatically as she left the building and heading for lunch in New Hall. Eyes on the ground in front of her – no need for an orientating scan of the now very familiar trek home – she prepared to reconstruct mentally the last lecture on mechanics and dynamics into a list of bullet points. Now, at the beginning of the penultimate week before the Christmas break, she felt at last able to acknowledge to herself the truth of an inkling that had taken root early in the semester and which had grown throughout the term to flower into full conviction – there was time in her schedule, after all, if she wanted it, for a boyfriend, just like normal people had.

Happily, her initial fears of not coping with the work had turned out to be groundless; on the contrary, she had found that, while she admittedly passed her evenings and weekends working, the time that she spent was in reading around the subjects rather than in trying to understand the lectures themselves, which she had found frankly unchallenging. This did not frustrate her, however, because she realised that the carefully structured lectures would give her the necessary foundations upon which to build the more exciting edifices of modern physics that she knew they would tackle in due course.

What she *had* learned during the first half of the semester was that the talent that she knew she had for getting to the root, the *essence*, of an equation or a description of physical phenomena, did not appear to be shared to the same degree by anyone else in her class, as far as she could see. She had largely maintained the low profile that she had resolved to keep, surfacing only when nobody else could come up with a suggestion for solving the problems that David liked to pose in his tutorials. It had become something of a ritual on these occasions for the class to turn round in their seats, mutely inviting Lucy to help them out, a ritual that was performed, on the whole, good-naturedly, with the possible exception of George who, as she had told David, had not bothered her after the question about the tunnel through the Earth.

The immediate significance of all this, though, was that, if she wished, she could have a life beyond the timetable, and enjoy herself – even with a boyfriend – without feeling guilty about not spending the time revising her lectures. Of course, in her mind, the concept of a boyfriend was in no way an abstraction, but was embodied in David, for whom her feelings had developed from an initial attraction, through to a fondness and eventually, now, into what she could only describe as love.

The question was, of course, what David thought of her. As far as she could see, he had no girlfriend – this she had concluded on the night of the fireworks party, because surely he would have taken a girlfriend if he had one. Of course, there was the chance that he did have a girlfriend but that she had been unable to come that night – unlikely, though. Nevertheless, she decided now that she would come right out and ask him when the right occasion arose, if she reckoned she could do it nonchalantly enough.

Lately, she had been trying to give him opportunities to reveal his thoughts, but perhaps she wasn't signalling strongly enough. She had expressed her concern for him at

260

different times over the Wilgoss affair – and she was, indeed, genuinely anxious for him – but, while he had been grateful for her solicitousness, he had not taken it any further. Of course, she could put that down to the very ghastliness of the whole business. She had missed an opportunity last Monday – the day of the suicide – when they had been in her room at the end of that awful day and she had led the way out of her room to show him back down to the entrance of the hall. If she had, instead, stood to let him out first, letting him brush past her, he might just have taken her in his arms…

Above all, though, she mustn't do the running. Oddly, that was a mantra that her Mum had said to her over and over again as she was growing up, even though there had only ever been the one platonic relationship she had had at school with her older boyfriend. But she reckoned it was good advice nonetheless. She would have to use her feminine wiles on David.

She began to draw up in her mind an alternative list of bullet points – this time featuring the reasons why she loved him. Looks, of course. He was tall and handsome. He was at the same time both protective and in need of protection, in a strange way. But high on the list was the way that they *connected*. They understood each other, not just in the trivial sense of understanding physics, but in a way that she had never before experienced. They had a similar sense of humour, they shared the same interests, their backgrounds were alike – their love of relativity, they had both lost parents recently, which made it easier to talk to each other about their bereavements – and now they were sharing experiences together, like the events surrounding Wilgoss and Paul Evans and writing their paper.

Contentedly distracted by these pleasurable reflections, she had just noticed that she had reached the shelter of the broad overhanging roof at the hall entrance and was reining in her umbrella when a man stepped in front of her and asked her if she was Lucy Darling.

'Yes, why, what's the matter?' she asked, immediately concerned for her Mum.

'No, nothing's the matter. My name's Kenny McGillivray – I'm a reporter with the *Courier*,' he said, naming the newspaper that Lucy knew was based in the city fifteen miles away but which served a wide region including the town and beyond. He held up a copy of *Stat* with her picture on it. 'I'd just like to ask you a few questions about the paper you were writing that Professor Wilgoss tried to hide. I gather the paper contained the key to the universe?'

'Sorry, no comment,' she said, already annoyed at the man for scaring her, however unwittingly, and enjoying her own obstinacy in retaliation, contrary to her nature though it was.

'Aw come on, pet, it's not a matter of national security.'

'I'm sorry, I said no comment.'

'Look, I've been to see David Lane and he gave me an interview and he said you'd give me one, too. So how about it?'

'Mr McGillivray, I understand you're just doing your job, and I'm sorry you've come all this way for nothing, but I said no comment.' With that, she dodged around him and walked smartly into the hall without looking back. Not until she had reached the dining room did she look round, but he had gone.

She felt a little ashamed with herself as she joined the lunch queue: it did not come naturally to her to be so brusque, but she had learned a lesson with Paul Evans. When she sat down at a table she pulled out her phone and called David.

'Hi, Lucy, how are you?'

She thrilled at hearing his voice right up in her ear.

'I've just had a visit from a reporter from the *Courier*. He said you said I'd give him an interview.'

262

'No, I didn't. I said *no comment* and he asked if I thought you would speak to him. I said you'd decide for yourself. He wanted to know where to find you, but I didn't tell him. He's obviously done some digging. What did you say to him?'

'Same as you – no comment.'

'Good, then, we've put up a united front.'

Lucy felt a glow of pride in not letting him down. 'I look forward to seeing you at the cathedral on Saturday, then,' she said.

'What about coming to my place for coffee first, to wake us up?'

Just try and stop me! 'That would be nice. What number is it?'

He told her and they fixed a time and she put away her phone to start her lunch, almost hugging herself with glee, so excited was she at the prospect of seeing David in his flat on Saturday.

Her satisfaction that she had done just the same as David with the reporter did not diminish as she munched her way through steak and kidney pie. She felt that she was really getting to know what David would do, and what he would say and how he would respond in a variety of situations. She supposed this must be what happened when you were in love – you began to know a person almost as well as you knew yourself. In fact, it now struck her that if she were to say that to David, he would smile and reply that she was building a model of himself in her brain. She toyed with the paradox – she had built a model of David in her mind, and this model was actually now saying to her that she had built a model of himself in her mind. *Weird or what?*

She wondered if David was thinking of her at that very moment, prompted, perhaps by the phone call. Had David built a model of herself in his mind? Well, he must have done. If only she could interrogate her model of him! So her model of him wasn't complete, or she would know all his

263

thoughts. Was this what the soul was – the fullest possible model of your mind, the essence of your being? Would it be blasphemy to think mechanistically like that if you were a Christian, or, indeed, a member of any of the organised monotheistic religions? She certainly didn't feel the thought was blasphemous, but then, she didn't hold with the dogma.

Lunch was over all too quickly, and she set off for the Physics Building, for her afternoon practicals. She was pleased to see no sign of the reporter. The rain had stopped, and the dripping campus sparkled in the golden light of the low December sun.

*

The bell in St Sally's tower was striking nine o'clock as she turned into Spey Street on Saturday morning. She began walking down the exclusive-looking, slightly forbidding street, lined on both sides with uniform rows of elegant, stone, Georgian terraced houses. The dwellings were on three floors – the ground floor, a first-storey level and a basement below street level, separated from the pavement by a void protected by black-painted railings, allowing daylight to reach down to the basement windows. The basement areas, which could be reached by separate flights of steps down into the light-wells, were presumably once the domain of the domestics and servants for each house and now, she guessed, served as self-contained flats at a lower rent than the rooms above.

She walked past a Victorian-style street lamp, found the house and crossed the stone slab over the light-well, rang the door bell and listened for the chimes. Almost immediately, David opened the door and greeted her with a big grin.

'Morning Lucy! How does it feel to be up on a Saturday when all the rest of the student population is still sleeping?'

'Actually, I'm feeling quite virtuous, getting up this early,' she said, following him into a large, high-ceilinged room lit by two floor-to-ceiling windows and decorated with a light-grey carpet, immaculately furnished with a three-piece suite upholstered in deep-red tufted velvet. There was no sign of the usual student detritus on the floor or on the sofa or armchairs – how could a bunch of students live so tidily?

'You take your coffee white with two sugars – right?' said David, hovering in the doorway after she had taken a seat, and then disappeared to prepare the drinks. It gave her a warm, snug feeling to think he was beginning to know her so well.

She was about to call through about how spick and span the flat was but stopped, not wanting to wake up the household. Instead, she started to examine the pictures on the walls. With a start, she realised that they were oil originals, not prints. There were several atmospheric views of stately full-rigged ships anchored at night in becalmed harbours, illuminated by moonshine and gaslight from houses on the harbour-side. Two other paintings were of rain-slick street scenes, again lit up by gas lamps reflected from the cobbled roads and pavements. She was bewitched by the paintings and rose from her seat to examine them more closely; she was not surprised to see they were all by the same artist.

She was just sitting down again when David reappeared carrying two mugs in one hand and a plate of ginger biscuits in another, setting them down on the coffee table between them.

'Are you still OK for the interview on Wednesday?' he asked.

She quickly refocused her thoughts. David had called her the previous day to ask whether she would like to do an interview with a reporter from the *New Scientist* magazine. The reporter had called him up after the *Courier* article had appeared under the headline *Death Prof hid God Equation*. The columns had contained no more information than the original story in *Stat*, except that they said the suicide had been connected with the professor's medical history and, crucially, the article reported the title of their first paper, which could only have come from Paul Evans' contact in the journal. The *New Scientist* reporter was only interested in the significance of the paper, not the suicide, and had asked David for an interview. David had replied that he would only grant one if his co-author was included, and had phoned Lucy to arrange an appointment.

'Yes, I'll be very interested to come,' she had said. 'But why are you happy to talk to the *New Scientist* but not to the other papers?'

'I know, it doesn't sound consistent on the face of it, does it? But the difference is that the *New Scientist* is a serious science magazine interested in the meaning of our papers, not the suicide. In fact, before I called you, I asked John Oakhill for his advice.'

'And he said go for it?'

'Not exactly. He said that, ideally, it would have been better to wait until the paper was published, but, since it was in the public domain now, what with the *Courier* and everything, it was probably better on balance to be interviewed than not, so we could kind of stake our claim.'

'So do we say anything about our next paper if they ask us?'

'Yes, I think that's probably best.'

With that, David had phoned back the *New Scientist* and they had agreed to meet on Wednesday afternoon.

They both reached down for their coffee, and Lucy picked up a biscuit.

'I hardly dare eat this,' she said, 'in case I drop a crumb on the carpet.'

'Oh, don't worry. To be honest with you, it's not usually as tidy as this. You provided the motivation for getting out the vacuum cleaner.'

'I'm still very impressed with the standard you people keep.'

David smiled. 'No, it's just me. Nobody else lives here.'

'What – you occupy the whole flat on your own?'

'Yes, well, actually, the whole house. The house is just for me.'

Lucy could hardly believe what she was hearing. 'You mean you own the whole house? The basement, too?'

'Yes, I do rattle around in it a bit. It was kind of an investment, I suppose.'

'But you didn't sell your house, did you? I mean in the west of Scotland.' *So where had he got the money?*

'Oh no, I wouldn't sell that. Actually, that's not just a house, it's an estate, a business. I know it sounds hackneyed, but my family have been there for generations.'

He was rich, then. He'd kept that quiet!

Elizabeth Bennett's joke, that she had begun to love Mr Darcy when she first saw his beautiful grounds at Pemberley, entered her mind unbidden. Well, she had begun to love David long before she knew he was well off. Anyway, she could hardly follow through, now, with her plan to ask him if he had a girlfriend!

'I was admiring your paintings,' she said. 'Did you take them over here from your estate?'

'No, the paintings there are just what you'd expect – hills, mountains, valleys, stags, all that sort of thing. These are all from the same artist who does wonderful things with light. I'm glad you like them.'

They discussed the paintings for a while and then David said: 'Talking of light, I had intended to give you the present of seeing the sun rise over the sea.'

'David, what a beautiful thought—'

267

He held up his hand, stopping her. 'No, it wouldn't have worked, sorry. I realised last Saturday when I saw the sun through the mist that, even if we had been there at sunrise, at about half past eight, it must be rising from behind the hills to the south-east just now. You'd need to be there in the spring or summer, when it rises to the east or even further north, so that it's rising over the sea.'

'Oh, never mind. That sounds like a good project for next year.'

'In the meantime, though, are you ready to go to Canon's Tower? The weather's just about perfect for a good view.'

*

Twenty minutes later, they stood at the base of Canon's Tower, having purchased two entry tokens. David led the way, pushing open the metal gate when his token released it. As they climbed the dimly lit spiral staircase, he called down to her, his voice echoing: 'You know, there's a story that the tower is supposed to have a ghost – a monk – and a student died because of it.'

'This sounds awfully like one of your White Lady stories!'

'The student was just as sceptical as you. He even bet that he would climb the tower at midnight to prove there was no ghost – this was in the nineteen-seventies, before they put the metal door on it. True to his word, he turned up at midnight. It was easy to open the old wooden door and he went all the way to the top, saw nothing, came back down and started to leave the tower. But just as he was leaving, he felt someone – or some*thing* – tug on his gown behind him – the students used to wear gowns all the time in those days. He dropped dead on the spot from a heart attack and was found on the ground next morning – the closing door had caught his gown as it flapped behind him and he died of the fright.'

Lucy was appalled. 'Oh my God! That's awful.' Then it dawned on her. 'You're having me on! You're making it up, right?'

David said nothing for a few steps and then whistled the haunting refrain from the *X Files*, the notes echoing eerily down the winding staircase, making Lucy laugh.

Finally they came to the last few steps and ducked out into the sunlight.

'Oh that's breathtaking!' she cried, gazing out over the parapet at the panorama of buildings spread out like a brightly cleaned model village beneath, the slanting sunlight highlighting roofs and etching sharp corners into the stone, rendering it crystalline, bestowing on the town a distinctly crisp, three-dimensional quality. 'Oh, and look at the harbour!' she exclaimed, turning round to see the breakwater punch its way far out to sea, that very jetty where she and David had walked and talked and sat in the sun looking at the first draft of their paper.

David seemed engrossed, too, although he must have seen the sight many times before. They spent quarter of an hour just staring at the view, Lucy occasionally pointing to a building or road she had just recognised. 'Everything looks so different from up here,' she said. 'It's as though we've come up to a higher meta-system and now we can see everything in a clearer perspective, and at last we can understand it.'

'Speaking of which,' said David, his tone, a curious combination of gravity and excitement, making Lucy look round at him, 'I think I know what Wilgoss meant by saying we might be surprised by what's in our paper.'

'You found something?' Lucy was thrilled. It was surely significant, because David had waited for this moment to tell her, here at the top of the world.

'Yes, and I'm almost scared to tell you what it is in case I'm deluding myself.' His excitement was now bubbling over, like a little boy with an enormous secret that he could barely contain. 'If I'm right, then no matter how huge you think this is going to be, you're underestimating it!'

'Come on, David,' she said, laughing, but impatient at the same time to hear what he had to say. 'Spit it out!'

'OK,' he said, visibly controlling himself. 'Think back to the Coffee House. Remember how you noticed that mathematical systems contain models of themselves – we called them self-similar in our paper?'

Lucy nodded.

'Well, I played around with the maths in our paper, and came up with this extraordinary result.'

'So what *have* you come up with?'

'OK, remember how we got to the God System – it's the end result of a whole hierarchy of meta-systems and meta-meta-systems and so on? Well, I wrote this down mathematically, and, as you might expect, there are two possibilities. One of them is that there is an infinite number of meta-systems, one above the other, which never leads to a final system. The other is that there is a number of meta-systems, one above the other, and the number might be infinite or not, but the meta-systems converge.'

'Yes, I get it,' said Lucy, 'Like adding one plus a half, plus a quarter, plus an eighth and so on forever. An infinite series but they simply add up to two point zero.'

'Yes, that's kind of what I mean.'

Lucy felt a pang of disappointment. 'But I thought we had more or less said this already? We agreed that the Inverse Gödel Theorem would no longer imply a higher meta-system if the meta-system contained generalized natural numbers, for instance, and so the series of meta-systems wouldn't necessarily go upwards forever.'

'No, no, you have to wait for the dénouement!' he cried, laughing.

'Sorry, go on.'

'Well, each meta-system that you meet on the way up will contain all of the mathematical systems below it, each with its associated version of the universe it generates – one inside the other, like the Russian dolls. Each meta-system, including its corresponding universe, contains the system and universe below it – a slightly incomplete model of itself, if you like – which is where the self-similarity comes in. The point is that you end up adding a smaller and smaller number of new axioms and rules in each higher meta-system as you go upwards towards the convergence.'

'Why, exactly?' she asked, reluctant to break his flow, but wishing to understand what he clearly thought was important.

'Because, in the higher meta-systems, the axioms and rules start being explained. You can't explain them in the system they apply to, but, in the higher systems, the reasons for them become evident.'

He searched around for an example. 'Like the axiom in what's called Peano arithmetic that says you can't have negative numbers. A higher system would allow negative numbers, and so in the higher system you can see that the choice of zero as the lowest number in Peano arithmetic is effectively arbitrary. But you couldn't have seen that in the Peano arithmetic system itself. Well, the axioms and rules gradually get mopped up as you get nearer to the God System.'

'So, as you get higher, are you saying that the systems become more and more general, allowing you different possibilities further down?'

'Yes, it seems that way. So, if you can picture it, as you rise up through the meta-systems with their associated meta-universes and you get closer to the God System, then, in any particular high meta-system at this high level, the model containing the systems below it is occupying just about all of this high meta-system – there are very few axioms and rules left that are not already proved in the lower systems – until, at the end, at the very highest level, the meta-system, including its meta-universe, is *completely filled* with the model – there are no remaining axioms and rules to add. So here's the dénouement – as the series of meta-systems and universes converges into the God System, you suddenly realise that the *model* in the final meta-system with its meta-universe is identical to the final system itself – the God System. *So the God System is completely self-referential – it is an absolutely complete model of itself – it explains everything!'*

Lucy couldn't speak; she just looked at him.

'Because it contains a complete description of itself – because it *is* a complete description of itself – it contains everything needed to recreate every mathematical system in the multiverse and beyond – *including itself*. It really *is* the God System!'

She found her voice. 'But surely something has to explain whatever rules and axioms remain in the God System?'

David gave an excited chuckle. 'I know, it does sound like pulling yourself up by tugging on your own shoe laces. Look, I thought of a crude analogy in our own mathematics that might help. You know that any well-behaved mathematical function can be broken down into an endless sum of its value at zero plus the slope of the function times its distance from zero, plus a half of the slope of the slope times its distance from zero squared and so on?'

'Yes. The Taylor series. Or, at least, the Maclaurin series.'

'Yes, that's it. Well the exponential function is very special here, because it happens to be its own slope, its own derivative. And, of course, it's also the derivative of its derivative, and so on. So the exponential function in a sense carries all the information needed to reconstruct itself – it's kind of a model of itself. Not a great analogy, of course, because, although the exponential function generates its own components, its own derivatives, it doesn't contain explicit instructions for putting them all together in a Maclaurin series, but it's the nearest I could come up with in our own primitive mathematics.'

'Yes, I see,' said Lucy, still dazed. 'So the God System would contain complete instructions for reproducing itself, a bit like the way DNA contains complete instructions for making a creature that makes more DNA?'

'Yes, that's an interesting analogy. But, of course, real DNA still relies on there being molecules and quarks making up those molecules, and the physical laws describing how they interact – ultimately, the mathematical system underpinning our universe. DNA doesn't supply these details. The God System does.'

'So, what do you think the mathematics of the God System would look like? What's the final God Equation?'

'Well, as you said, it would have to be very general. It might be as bland and uninteresting as – what – a sphere, say. But, if you split open that sphere into two pieces, not neatly cleaved but leaving ragged edges, then each ragged-edged piece becomes interesting. You can't easily explain one without referring to the matching edge in the other half of the sphere.'

Lucy was galvanised. 'David, you need to get this published as soon as possible. Before someone else reads our paper and comes to the same conclusion.'

'I know. I've had a bash at writing a draft already. It was frighteningly easy to write: I mean, it was so straightforward that anybody will be able to do it once they read our paper. I was thinking I'd ask for John Oakhill's help to see if he could smooth its passage in the circumstances. Apparently, he knows the editor of *Int J Nat Phil*. I was hoping you'd be able to have a look at the draft and then I could send it off. What do you think?'

'I think the sooner the better. Could I look at it today? This morning?' That would still give her time to go with Gillian to visit poor Sarah, who had undergone a thyroidectomy in the city hospital the day before. She would be an inpatient for a few days, including over the weekend.

'Thank you, Lucy. I was hoping you'd say that. Shall we go back to my place now, then?'

'Yes, that would be a good idea. I'll have to leave about half past one to visit my friend in hospital. She had an operation yesterday. I'm going with another of her friends.'

They clumped down the spiral staircase faster than they had climbed it and set off for David's house once more.

Thinking about the consequences of David's remarkable result, Lucy began to pursue another line of thought as they started down South Street. *Dare she ask him?*

'David, you've always said,' she ventured slowly, 'that self-awareness in a being arises through having a model of itself in its mind. So could the God System be self-aware? If it were, then it would be *completely* self-aware, wouldn't it, given that it's a perfect model of itself?'

'Sorry, Lucy. I believe you're right, that a system has to be self-referential – it has to contain a model of itself – as a *necessary* condition for it to be self-aware. But it's not the only condition – otherwise, as we've said before, the nest of Russian dolls would be conscious.'

'So, what's the missing magic ingredient for self-awareness?'

'Well, for one thing, the model can't be static – it can't just be the painted face of a doll, for example. The model has to change to reflect new data that is discovered about the environment and itself – it has to be continually updated. And there has to be a part of the model that can experiment with possible changes to that environment and itself. But one of the difficulties with the idea of the God System being self-aware – or any meta-system and its meta-universe, for that matter – is that the idea of *updating* and *changing* doesn't seem to make sense for a meta-universe in which the dimension of time hasn't been generated. You can't have change, and you can't have updating, without the dimension of time, or its equivalent, in the meta-universe. Sure, time has been generated in our own universe, but it's not clear to me that its equivalent is similarly generated in the higher levels.'

'OK, I see the argument. But it seems strange to me to think that there should be pockets of self-awareness in our own, imperfect universe and yet that this special quality should be lost further up. After all, didn't you say that everything in our universal mathematical system and the universe itself would be accessible – *visible*, in a manner of speaking – to the higher meta-systems and meta-universes?'

'You should join the debating society – you present a well argued case! One thing we can surely agree upon is that the God System permits self-awareness in our universe. Maybe the difficulty is in defining self-awareness at higher levels in universes where there is no distinct dimension of time – where all possible scenarios, perhaps, are already played out to their ultimate conclusion. I think we'll just need to keep this one simmering, and I promise I'll keep an open mind on it.'

'That's fair enough. So, do you think Wilgoss saw what you've seen, then?' she asked, changing the subject.

'Pretty much, I should think. The equations are there in our paper – once you read the draft, you'll see how small a step it is to follow the hierarchy of meta-systems and converge at a completely self-referential one.'

'And Wilgoss equated this with God?'

'No, his religious beliefs wouldn't let him. But he reckoned that others might. That's why he tried to stop the paper.'

'So, he thought this was – what – a *threat* to conventional Christianity?'

'I wouldn't put it as strongly as that. My guess is that the majority of Christians wouldn't be influenced in their belief in their god by mathematical or scientific evidence, even if, as in this case, the evidence tended to support aspects of their belief. But I think a hypothesis that deduces mathematically a system that, effectively, is the source of everything in the universe and the multiverse and beyond, *including itself*, would be an uncomfortable fact to live with, for those Christians who understood it. For that matter, it might be uncomfortable for atheists, too,' he added with a smile, looking at Lucy.

'I hope you appreciate the irony – you, an atheist, coming up with this,' she said.

'Well, the intellectual exercise has been great fun – it was a genuine *Eureka* moment when the completely self-referential solution popped out of the equations – but, of course, I'm uncomfortable with the interpretation that some will put on it – that the God System is actually a god. Ironic, when you think that Wilgoss was saying the same thing from the other side of the fence!'

'I suppose, in a way,' said Lucy, 'that's a good test of the paper. If a dyed-in-the-wool atheist deduces it, then that has to earn it some street-cred!'

They walked on in silence for some time, each deep in thought, turning automatically together into Church Street and Market Street.

'The million-dollar question, of course,' said Lucy eventually, 'is whether you're still actually an atheist!'

'Hah! I was wondering if you'd ask me that! Look, here's how I see it. Like the majority of the human race, I wonder about what led to the creation of the universe. Many physicists will be satisfied with an explanation about the big bang, or colliding branes or bouncing universes or whatever, but I want to know what caused the laws and the mathematics that underpin the universe and which led to the big bang or the branes in the first place. Many people,

276

including some physicists, say that a god must have created the laws and the mathematics, and they endow their god with a personality. For me, that approach invites at least as many questions as the answers that it supplies – it's too glib. However, that still left me with the central question.'

He paused while they separated to let a mother holding her young child's hand pass between them.

'Then you came along,' he continued, 'with the idea of an Inverse Gödel Theorem that points to a hierarchy of meta-systems beyond that of our own universe. We follow the logic and find that either the hierarchy goes off to infinity, which is just as dissatisfying and is anyway unnecessary, or it converges to a very special system that we nicknamed the *God System*, which turns out to have the property that it is a complete model of itself – it explains itself, and every system below it, including the multiverse, if it exists, and our own universe for good measure. For me, that's a satisfying explanation of existence. It's not what most people call *God*, though.'

'So you believe in a system that created everything, including, in a way that I don't quite understand yet, itself. Some might call that *God*.'

'Well, yes, that's the dilemma for me. Being an atheist has kind of defined me for much of my life. Maybe we need a word – a *metacosmist* – for a person who concludes from scientific principles that there is a higher level to the universe. That would separate me from followers of the dogmatic religions! But what about you? How would this paper affect you?'

'Ever since I could think for myself, I've felt that God must exist because, otherwise, you end up with an infinity of questions. Every time physicists think they've explained the universe – the big bang, or the quantum fluctuation that maybe caused it, or the collision of branes that you refer to, then you have to find an explanation for *that* phenomenon. To me, the only way to get closure on this infinite staircase

277

of questions is to top them off with God. So, to answer your question, I'll be delighted if the logic of your draft is correct, because it's a scientific, a mathematical way to express what my gut feeling has told me ever since I began to be old enough to question these things for myself.'

Turning into Spey Street, they arrived at David's house and he led the way into the drawing room where they had sat earlier in the morning. David left and returned with a printout of his draft paper. They sat side-by-side on the luxurious sofa and he began to go through the manuscript with her.

By one o'clock, they had finished, and she followed David through to the kitchen to eat a pizza delivered by a local service. They sat down on stools beside a central island topped in black granite – sensibly, this one, unlike those she had seen on television, had a generous overhang, so that you could get your knees under the top.

'I'm surprised to see such a large kitchen in a house that must go back a couple of centuries,' she said.

David uncorked a bottle of Valpolicella. 'Oh, this room was converted quite recently. The original kitchen was downstairs, in the basement. I didn't do the conversion – in fact, I'm ashamed to say I've done practically nothing to the house since I moved in.'

'How long have you had it? Cheers!'

'Cheers! I bought it after I graduated. I saw it as kind of an investment – my idea was to live in it while I did my PhD and then, if I moved on, I could sell it.'

So he had enough money to buy this place even before his father died, she thought. She had assumed that he had inherited everything when his father died – what had he said? – a year ago.

'But you're hoping to do a post-doctoral fellowship here?'

'Yes, I hadn't thought that far ahead when I bought it, but I've grown to love this town, and the university and its people.'

So have I! 'So what about your estate?'

'Ideally, what I'd like would be to do my work here, where I'm rubbing shoulders with other physicists and I can go to meetings and conferences and what have you, and to go back to the estate as a kind of retreat for the summer holidays and Christmas and so on. But it gets a little lonely up there after a while, despite the visitors.'

'What visitors?' she asked.

'Oh, the estate is managed by a specialist firm. They arrange corporate entertainment, you know, clay-pigeon shooting, archery, mountain-biking, that sort of thing, so we get quite a few visitors. It provides an income and keeps the estate ticking over. They're completely trustworthy, the estate managers: they've worked with my family for many years. They want to develop it further, but I'm not sure about that at the moment. We'll see.'

He took a gulp of wine. 'So, what do you think about the paper?'

'I suppose my first impression is that I'm surprised at how short it is. It's succinct. But, technically, as far as I can follow the logic, it seems to hold together. The very last equation looks almost holy – there's so much meaning packed into it. *I am that I am.*'

'Sorry?'

'That was supposed to be what God said to Moses when he asked for His name. Quite self-referential, don't you think?'

'Goodness, Lucy, that's creepy! If I didn't know the Bible was written by human beings, I'd say that's a pretty direct translation of the equation!' He took another sip from his glass, and recalled the analogy that he had tried on Christos in the Physics coffee room. This time, he would add a refinement. 'I suppose another way to visualise what's

279

going on would be to think of a television screen as a mathematical system. Now suppose I have a camera and the screen shows what my camera is pointing at. If I point my camera at the TV, the screen will show the room with the TV in the corner.'

'So what you have is a model of the mathematical system displayed on the screen – so the mathematical system has a model of itself,' said Lucy.

'Exactly. But there's more in the mathematical system than just the model, of course – there's everything you can see around the TV set in the room. That can represent the mathematics that is at a higher level than the mathematics in the model. But now suppose that I start to zoom in on the TV set. Then the TV will occupy a greater and greater area of the screen.'

'So, when you zoom in,' said Lucy, able now to follow the argument more easily, having just read the paper, 'you're doing the equivalent of looking at one of the higher meta-systems, as you get close to convergence with the God System.'

'Absolutely! As you zoom in, there's less and less surrounding the TV set, and that corresponds to the fact that most of the mathematics in the higher meta-systems is already in the lower systems – it's already in the model. Finally, as you converge on the God System itself, you zoom in so much that only the screen itself is visible in the image of the TV set. You can't see the frame of the screen any more – only the screen itself. So now, the meta-system is showing only the image of itself, and nothing else. If you lean forward with a magnifying glass, you'll see the individual pixels on the image corresponding exactly to the physical pixels on the screen in front of you. The meta-system is now exactly its own model!'

'That's a good analogy,' said Lucy. 'Thank you.'

'Of course, it's far from perfect, but I think it gives the flavour.' He glanced at his watch. 'Lucy, the last thing I want to do is to cut short your time here, but do you realise it's a quarter past one?'

'Thank you, David. I don't want to leave, either. It's been a brilliant morning – I really mean that. But I have to go: Sarah's depending on us.'

'Thank you, Lucy, as ever, for helping me out and checking the draft. I've got the confidence now to see John Oakhill on Monday and hope he'll have some ideas on speeding up the publication process.'

And so she set off to meet Gillian at the bus station round the corner, her heart singing.

CHAPTER 19

Wednesday 15 December

David needed no psychology book, no teenager's magazine to tell him he was in love, but he had noted, with an amused interest, that he was displaying all of the symptoms that would have been listed in either or both of these sources as conclusive evidence for the syndrome. Theoretical physics was an especially unfortunate occupation for those afflicted with the condition, because it involved no physical activities that could count towards a tally of work accomplished: progress in the subject was effected only by sitting and thinking deeply, and that, of course, was an open invitation for the love bug to strike, to take control of his ruminations, to nudge his appreciation of the beauty of the equations he was dreaming up, into a celebration of the beauty of the object of his love. So he had achieved nothing all that Wednesday morning, and, as lunchtime and his appointed meeting with Lucy approached, he gave up all pretence of writing any more of his thesis for the day.

He had arranged to meet her after her morning lectures and have lunch in New Hall, Wednesday afternoons being free time for her. They would then be interviewed by the *New Scientist* reporter. After checking that Lucy liked seafood, he had booked a table for them both later that evening in a sumptuous glass-walled restaurant perched on rocks and overhanging the sea. Of course, the view, stunning in daylight, would be lost to them at that time of day, but, if the food were even half as good as its reputation, then the meal would be unforgettable. Being on one edge of the town, it would be accessible on foot and so he needn't worry about drinking and driving.

The reason for the dinner, which, he had made clear to Lucy, would be his treat, was ostensibly to celebrate the completion of their two papers. However, he had another motive for arranging the outing. Completion of the project would remove the pretext, if one were necessary, for meeting her regularly. He simply did not know what weight she had placed on the project itself for meeting him and how much she did so because she liked his company. He considered himself a reasonably sensitive person, but, probably through his lack of experience with women, he had been unable to interpret with confidence the signals, if any, that she had been giving him over the last ten weeks. He had no doubt that the differences in age and stage of career might have been an issue, at least at the beginning of their unusual relationship, but he hoped that they had come to know each other sufficiently well that such differences no longer mattered.

While the project had been running, he had chosen not to risk admitting his growing fondness, and, indeed, love, for her, for fear of scaring her off. Now, however, with the project finishing and, moreover, the prospect of the two-week Christmas break looming, he faced a future without her presence, except remotely in tutorials, which would be an unbearable agony.

He therefore reasoned that he had nothing to lose by using the occasion of the dinner, or perhaps the walk back with her to New Hall afterwards, to ask her out on a date next term. *For old times' sake*, he might put it – he couldn't quite bring himself to put all his cards on the table, even now, at this late date.

He had met with John Oakhill on Monday morning and had given him the draft manuscript, along with the paper that had already been accepted. Oakhill had very kindly set aside a couple of hours there and then so that he could go through the papers with David. By the time he had come to the last page of the manuscript, his eyes were shining brightly.

'What *have* you done?' he asked, turning to David, his face a mixture of surprise, astonishment and excitement, his mouth open in a gaping smile.

David couldn't quite read him, and so he opted for a safe, neutral reply. 'What do you think of it then?'

Oakhill spread his hands out, palms upward, the same expression on his face, shaking his head slightly. 'Words fail me!'

'Do you see why I'm concerned that, once the news of the first paper is out, then the results of the second paper might be obvious?'

'Aye, I do. That's the beauty of it – it's obvious once you see it. But you're right. You need to get this published straight away.'

So, there in his office, Oakhill had lifted his phone and called the editor of the *International Journal of Natural Philosophy*, his old friend from undergraduate days at Cambridge. In ten minutes, it was arranged that David would submit in the usual way and then email the editor with the reference number of the paper, and the editor would ask the same referees on the editorial board to look at it. David was, of course, delighted.

'Don't think it will always be this fast,' said Oakhill with a wry smile. 'You'll take badly the next time you submit a paper.'

'Thank you, John. I owe you, big time.'

David got up to return to his room and submit the paper. As he reached the door, Oakhill said to him, 'You know, David, before I read your two papers, I was an agnostic, but now I really don't know!' and with that, they had both burst out laughing, the tears finally rolling down Oakhill's cheeks.

*

Lucy came out of the lecture theatre with her fellow students at five to one, her morning's lectures over, looking for David. As always, now, when she saw him waiting there for her, her heart jumped, and she couldn't keep the grin from her face, delighted to see David grinning in response.

'How was the lecture?' he asked her, as they walked to the entrance and out towards New Hall.

'Pretty intense, actually,' she said. 'I thought they would wind down towards the last day.'

'No, it stays pretty steady right to the end.' A thought struck him. 'Do you realise I'll be giving the last formal teaching slot before the Christmas break?'

He was right: the tutorial on Friday at two. 'Well, you can count on me being there,' she said, saddened to hear him talk of the imminent break, wishing that she dared say more, hoping that David could read her.

David asked her how Sarah was. He was always a little reticent when it came to asking about women who were ill, in case they turned out to have some sort of gynaecological problem. To his relief, he learned that the operation had been on her neck.

Over lunch, he enquired about Lucy's mother. 'She must really be looking forward to seeing you,' he said. 'It must have been a long three months for her, alone without you.'

'She hasn't admitted as much, but you're right, of course,' she said. 'I phone her regularly, but probably not as often as I should.' As in many previous conversations with David, she thought how sensitive, how thoughtful and considerate of other people, he was.

David allowed himself to fantasize for a moment. He projected himself in the restaurant that evening, finally daring to declare his love for Lucy at the end of their meal, the candles on the tables around them multiply reflected in the dark glass windows, mellowing the golden Sauternes

285

with which they had washed down a dessert of prune and Armagnac pudding. She would have responded that she loved him just as he loved her and then they would make plans not to be apart over the Christmas break; they would plan for their future. He would drive down to Southampton with her and stay with her and her mother and then the three of them would go up to his estate where they would greet the New Year.

As Lucy ate, she tried not to stare at him as he sat opposite, tried not to eat him up with her eyes. Even sitting down, he looked tall, handsome, special: he stood out among the other students. She felt possessively proud of him: to the other diners, they would clearly belong to each other – he would not be recognised as being from the hall and so he must be there because of her. She wondered what he would do over the Christmas holidays. Would he stay at the university or visit his estate? She had pictured his estate repeatedly in her mind since Saturday, embellishing it with further detail with each mental visit. There would be snow on the ground and on the branches of the pine trees beyond the clearing. This time, she was part of the picture, and she and David would be sitting in the opposite corners of a chintz-cushioned window seat, looking out over the snow-clad lawn, perhaps at a deer, emboldened by the search for food to venture onto the clearing. They would be sitting in the huge drawing room with their backs to a great log fire, the smell of wood-smoke unearthing nostalgic memories of her childhood. Suddenly, she felt selfish. What about her Mum? She would need to update her scenario and include her next time.

They finished their lunch and walked back to David's office in the Physics Building, the sky quite dark, although it was only two o'clock, threatening an imminent downpour, or perhaps snow. David was just about to make coffee when John Oakhill knocked and came in.

'Sorry, David,' he said, seeing he had company, 'but I've got some news you'll both want to hear.'

'Goodness – what is it? But would you like some coffee, John? I'm just about to make it.'

'No, I won't thanks, I've just had mine. What I came to say is that Colin Howarth – the *Int J Nat Phil* editor – just phoned me to give me a heads-up. Both referees are going to recommend publication of your paper.'

'Oh, that's brilliant, thank you, John. And thank you for taking it personally to Colin Howarth. Still, that's very fast, isn't it?'

'Indecently so! Apparently they were both bowled over by it. Oh, and before I forget, they're recommending that you call the two papers parts one and two. I don't see any problem in that – all you need to do is to reply to the email that Colin will be sending you if you agree.'

Lucy added her thanks to David's and Oakhill left, just as the office phone rang. It was Ann, the receptionist, saying that David's visitor had arrived.

David took Lucy with him down to Reception, but they could see who their visitor was without having to ask. Standing in the foyer was a smartly dressed woman of about forty in a dark powder-blue woollen suit with a knee-length skirt: clearly not a student. She held out her hand to introduce herself.

'Hi, I'm Debbie Russell from the *New Scientist*. You must be David Lane, and you must be Lucy Darling.'

They shook hands and David led the way up to his office.

While David prepared coffee, Debbie handed them both her business card and the latest issue of *New Scientist*. David glanced at the card and saw she was both a writer and consultant for the magazine. He did a double-take when he noticed a PhD discreetly appended to her name in a small font.

'Thank you for agreeing to meet me at such short notice,' she said. 'I expect you're fed up with the media crowding you.'

'Only the university rag and the local newspaper, so far,' said David. 'But I think those experiences taught us a lesson – we won't be giving interviews to any paper that would use Professor Wilgoss's suicide to sensationalise the story.'

'Yes, it must have been a horrid time for you,' said Debbie. 'We picked up the story on the web, of course, but it wasn't the Wilgoss aspect that attracted me: it's the back-story about your mathematics paper that interests me. When I saw the title of your paper in the *Courier*, and then you told me on the phone that it was based on Gödel's Incompleteness Theorem, I knew I just had to follow this up.'

'Dr Russell—' David began, causing Lucy to jerk her head down to look at the card again.

'Oh, please, call me Debbie. '*Dr Russell* sounds like crinkly dried leaves!'

'OK, Debbie,' said David, 'we'll be happy to tell you about both of our papers, but can you tell me – is your PhD in physics or mathematics?'

'There are *two* papers?' She arched her eyebrows in surprise. 'Yes, the PhD is in cosmology, so think theoretical physics. That's why I'm particularly interested in your work. It will be my job to interpret and present what you say so that it can be broadly understood by scientists in all of the different fields represented within our readership, as well as the many who are not practising, professional scientists at all, but are interested in science all the same.'

She took a sip of coffee. 'So, if you'd like to take me through it from the beginning – what would you say was the germ of the idea for your first paper?'

And so they talked with Debbie for three hours, David first explaining how he had been stuck up a blind alley with Gödel's theorem until Lucy had had the inspiration to turn

288

the theorem on its head, and how this had pointed to a higher mathematical meta-system that solved the Gödel sentence expressed in any given mathematical system below it. Lucy had noted that each meta-system looked similar to the meta-systems above and below it, in that each contained a description, or a model, of the mathematical system below it. This self-similarity would have important consequences for the second paper.

Then they went on to tell Debbie how, in the second paper, they had applied these ideas to the mathematical system that generates the universe, resulting in an ascending hierarchy of meta-systems along with the corresponding versions of meta-universes that they generated. There were two possibilities – either that the ascending staircase of meta-systems and associated meta-universes went on for infinity, which turned out to be mathematically unnecessary, or that the staircase eventually levelled out into a mathematical system and its universe which turned out, intriguingly, to be a complete model of itself, a complete prescription for generating all the meta-systems and universes below it, including itself.

'And that would be the famous God Equation!' Debbie remarked.

'Well, to be honest, when the *Courier* came up with that phrase,' said David, 'we hadn't seen all of the implications of our first paper at that point. Oh, look, it's snowing.'

Sure enough, they had been so absorbed that nobody had noticed. It was by now too dark to see outside, but a thick layer of snow was already lying on the window ledge, covering the bottom of the window pane. David got up, put his face to the window and cupped his hands round his eyes to cut out the light from the office, leaving the lights of the quadrangle to illuminate the substantial covering of snow that had fallen during the afternoon.

'It's very thick out there. I hope you've got reasonable transport, Debbie.'

Debbie assured him that her 4x4 was up to the task. She was staying in Edinburgh that night.

'For what it's worth, Lucy, David,' she said, rising from her chair, 'I've covered many stories in my time as a journalist, but I've never, ever, come across one like this.'

Lucy's heart sank: this street-wise woman from the real world was about to burst their bubble.

Debbie caught her eye and laughed kindly. 'Don't look so dismayed, Lucy,' she said. 'I was about to tell you that this is the most exciting story I've ever been lucky enough to cover!'

They looked at her, their eyebrows questioning her.

'I've reported on all sorts of theories in cosmology. You know yourselves how many different scenarios have been proposed to explain our universe. Some of them are very weird, and many of them you can see right away involve special pleading. All of them raise as many questions as they purport to solve. If branes collide, then what causes the collision? Why are they waving around? If there is a multiverse, then where did the laws arise that permit a multiverse? And so on. Of course, your papers go beyond the particulars of the creation of our universe, although, eventually, it may be possible to start from the God Equation and work downwards to derive some testable propositions. But what is so different about your work is that it feels *right*. To me, it feels like coming home! And the beauty of it is, it doesn't leave an unanswered question – the answers are modelled inside the God Equation!'

Lucy and David looked at each other, couldn't help grinning at each other.

'So what's your next move? Will you try to find the characteristics of the God Equation?' Debbie asked.

Of course! Why had he not thought of that! Their time together on the project was not over by any means, David realised. Just as he had discussed with Lucy, there was nothing preventing anyone from writing down the axioms and rules of a higher system in this universe – they would, of course, be unprovable in this universe. But the signature characteristic of it being a model of itself would be unmistakable. So maybe that's where they had to start!

He kept his voice level. 'Yes, that could be our next project, couldn't it, Lucy?'

Lucy was ecstatic. This wouldn't be the end of their working together after all! She smiled and replied in the same tone, 'Yes, that makes sense, David. I'm up for it. Should we tell her about the self-awareness thing?' she added.

'Go on then,' said David.

'What self-awareness thing?' asked Debbie.

So Lucy first explained how David believed that, for self-awareness, you needed to have in your mind, continually updated, a model that included not only your environment, but a model of your own mind also.

'Are you saying the God Equation would be self-aware – aware of itself?' Debbie was incredulous.

'Well, David and I differ on that. But he still has to come up with a convincing argument why the God Equation, or its associated meta-universe, isn't self-aware in some sense. It may not ever be possible for us to have more than an inkling of what such self-awareness could mean. It's maybe something else we can work on?' she said, turning to David, who just smiled back, enigmatically, and winked.

'I'll have to think about that,' said Debbie. 'I might put that in, and I might not. Either way, your work has obvious religious significance.'

A thought seemed to strike her and she added, 'has this changed either of your attitudes in the religious sense?'

Lucy spoke first. 'Yes, definitely for me. I believed in a god before – not any of the dogma gods, but in *something*. Now I feel I have a much surer foundation for that belief, and I would say that's a comfort to me.' She looked at David for him to reply.

'Like Lucy,' he said, 'I have to say that our work has changed my views as well – perhaps even more than Lucy's. I was dogmatic before, not in the religious sense, but as an atheist. But I see now that I was fighting a straw man – I was fighting for the corner that said there was no god like the Christian god, or, indeed, any of the gods of the formal monotheistic religions. So Lucy and I were of the same mind in that sense. But my mind was just as affected by dogma as the religions that I decried. Now, I can see that there are possibilities so far beyond that severely limited concept of a god that I feel very humble in the light of Lucy's wisdom.'

None of them spoke for a moment. They stood looking at each other. Lucy's heart was almost bursting with pride, not for herself, but for David.

Debbie broke the silence. 'Well, if I want to get there tonight, I'd better get a move on,' she said, and they accompanied her downstairs into the foyer, Debbie promising to show them a draft of her article before it appeared in print, which, she said, would be in about a month.

'I feel quite drained,' said David, as they watched her tail lights disappear round the corner into the darkness. 'It's stopped snowing – do you fancy a walk before we go to the restaurant?'

And so they went back to his office where they put on their anoraks and scarves and set out towards the town. Their shoes scuffed into the sparkling snow and sent little puffs of diamond-dust into the air with each step. The night sky was clear now, the high December moon riding in a halo of pearly grey, outshining the stars and bathing in its cold light the freshly fallen fluffs of snow, blanketing roofs, roads and pavements, edging twigs and branches with silver and transforming the town into a monochromic Christmas card.

As they entered Market Square, they were greeted with the voices of carol singers ranged around a Christmas tree provided by the Merchants' Association. The square was lit by the blue and white lights on the tree and by the multi-coloured illuminations surrounding the square. With surprise, David realised that he no longer felt guilty about enjoying the sound of religious singing – he no longer felt defensive.

'It's beautiful, isn't it?' said Lucy, taking in the snow and the lights and the singing, folding her arms against the sub-zero temperature.

'Yes, it is beautiful.' Seeing her shiver in the corner of his eye, without thinking, intoxicated by the moment, David threw a protective arm around Lucy's shoulders to shield her from the cold, and then, horrified at being so forward, kept it there, afraid that to withdraw it would draw more attention to his gaffe.

To his amazement, he felt her arm come round his waist. He tightened his grip and she responded by squeezing him more. They stood like that, motionless, for sweet, timeless heartbeats, neither daring to break the spell.

'What are you doing for the Christmas break?' Lucy asked him, eventually.

Lightning Source UK Ltd.
Milton Keynes UK
07 January 2011
165311UK00001B/6/P